Max-Planck-Institut für ausländisches
öffentliches Recht und Völkerrecht

Beiträge zum ausländischen öffentlichen Recht und Völkerrecht

Begründet von Viktor Bruns

Herausgegeben von
Armin von Bogdandy · Rüdiger Wolfrum

Band 228

Matthias Ruffert · Sebastian Steinecke

The Global Administrative Law of Science

in co-operation with Jana Mühlisch

Springer
the language of science

ISSN 0172-4770
ISBN 978-3-642-21358-8 ISBN 978-3-642-21359-5 (eBook)
DOI 10.1007/978-3-642-21359-5
Springer HeidelbergäDordrecht London New York

Library of Congress Control Number: 2011931675

Die Deutsche Nationalbibliothek verzeichnet diese Publikation in der Deutschen Nationalbibliografie; detaillierte bibliografische Daten sind im Internet über http://dnb.d-nb.de abrufbar.

Einbandentwurf: WMXDesign GmbH, Heidelberg

Gedruckt auf säurefreiem Papier

Springer ist Teil der Fachverlagsgruppe Springer Science+Business Media (www.springer.com)

Table of Contents

Acknowledgements

This book is the result of a research project funded by the German Research Foundation (Deutsche Forschungsgemeinschaft, DFG), which also supported its publication. We wish to express our gratitude towards the DFG for this assistance. Further, we wish to thank the student assistants subsequently active in the project: *Maria Busse, Carolin Damm, Katja Frey, Ulrike Pollin, Luise Schöne, Michael Sellner, Eric Urzowski.* Special thanks go to *Katrin Rentzsch* who contributed to Part D. IV. and considerably supported the proofreading and, of course, to *Susanne Prater* (secretary) who was in charge of the final editorial work.

"Le savant a une patrie, la science n'en a pas."
Louis Pasteur[1]

Introduction: Science as a Field of Research for International Law

Astonishment could not have been greater particularly outside the scientific world, when in two articles in the journal "Science" of 2004 and 2005, the South Korean veterinarian (!) researcher *Hwang Woo-Suk* reported to have succeeded in cloning human embryonic stem cells[2]. Should a long race in biotechnology devouring massive resources have finally found a "winner"? Should there be a biotechnological solution to end such plagues heavily burdening mankind such as cancer and aids or such afflictions as Parkinson's and Alzheimer's disease, stroke, arthritis, diabetes, burns, and spinal cord damage[3], should human organs be replaceable – but also: should this, above all, be a further step in man becoming the creator of himself? Astonishment turned into shocked disgust when in 2006 *Hwang Woo-Suk's* "research" was revealed to be the result of fabricated experiments. The scientific publications had to be revoked, *Hwang Woo-Suk* lost his post as a university professor and had to face criminal proceedings, resulting in him being found guilty of embezzlement of enormous sums of money and sentenced to two years

1 Louis Pasteur, *Inauguration de l'Institut Pasteur, Annales de l'Institut Pasteur*, 1888, pp. 29 *et seq*. – quoted from Robert Merton, *Social theory and social structure*, 1968, p. 608.

2 Hwang Woo-Suk et al., "Evidence of a Pluripotent Human Embryonic Stem Cell Line Derived from a Cloned Blastocyst", *Science* 303 (2004), pp. 1669-1674; Hwang Woo-Suk et al., "Patient-Specific Embryonic Stem Cells Derived from Human SCNT Blastocysts", *Science* 308 (2005), pp. 1777-1783. Editorial retraction of these papers: *Science* 311 (2006), p. 335.

3 James A. Thomson et al., "Embryonic Stem Cell Lines Derived from Human Blastocysts", *Science* 282 (1998), pp. 1145-1147 at pp. 1146 *et seq*.; Constance Holden and Gretchen Vogel, "Cell Biology: A Seismic Shift for Stem Cell Research", *Science* 319 (2008), pp. 560-563 at pp. 560 *et seq*.

suspended imprisonment by the Seoul Central District Court in October 2009[4].

We live in a world of science. Scientific progress, the knowledge-based society[5], economic performance driven by innovations and ethical boundaries to research are only a few widespread keywords underlining this platitude. It goes without saying that the world of science is a borderless world – la science n'a pas de patrie. Therefore, the scandalous case of *Hwang Woo-Suk* remarkably illustrates the legal dimension of a borderless world of science[6]. As noted above, the perpetrator of fraudu-

---

4 Cf. Péter Kakuk, "The Legacy of the Hwang Case: Research Misconduct in Biosciences", *Science and Engineering Ethics* 15 (2009), pp. 545-562 at p. 546; David Cyranoski, "Woo Suk Hwang convicted, but not of fraud", *Nature* 461 (27 October 2009), p. 1181; *Zeit-Online* 26 October 2009 "Genetiker Hwang Woo Suk verurteilt".

5 The notion of the "knowledge society" was first used in the late 1960s in the works of Robert Lane, Peter Drucker and Daniel Bell: Robert Lane, "The decline of politics and ideology in a knowledgeable society", *American sociological review* 31 (1966), pp. 649-662; Peter Drucker, *The age of discontinuity: guidelines to our changing society*, 1969; Daniel Bell, *The coming of post-industrial society: A venture in social forecasting*, 1973) and taken up by Nico Stehr in the early 1990s: Nico Stehr, "Modern societies as knowledge societies", in: George Ritzer and Barry Smart (eds.), *Handbook of social theory*, 2001, pp. 494-508). For the later development cf. Deutscher Bundestag (ed.), *Schlussbericht der Enquete-Kommission Globalisierung der Weltwirtschaft – Herausforderungen und Antworten*, Drucksache 14/9200, 2002; Martin Heidenreich, "Die Debatte um die Wissensgesellschaft", in: Stefan Böschen and Ingo Schulz-Schaeffer (eds.), *Wissenschaft in der Wissensgesellschaft*, 2003, pp. 25-51; Nico Stehr, *Wissen und Wirtschaften. Die gesellschaftlichen Grundlagen der modernen Ökonomie*, 2001; Manuel Castells, *The Rise of the Network Society. The Information Age: Economic Society and Culture*, Vol. 1, 1996, Joachim Bischoff, *Mythen der New Economy. Zur politischen Ökonomie der Wissensgesellschaft*, 2001; Helga Nowotny/Peter Scott/Michael Gibbons, *Re-Thinking Science: Knowledge and the Public in an Age of Uncertainty*, 2001. Further Helmut Willke, *Dystopia*, 2002; Rolf Kreibich, *Die Wissenschaftsgesellschaft*, 2nd ed. 1986.

6 On the internationalisation of science in general see Wissenschaftsrat, *Empfehlungen zur deutschen Wissenschaftspolitik im Europäischen Forschungsraum*, 2010 (Drucksache 9866-10), at pp. 18 *et seq*. Cf. for an earlier assessment Vittorio Ancarani, "Globalizing the World – Science and Technology in International Relations", in: Sheila Jasanoff/Gerald E. Markle/James C. Petersen/Trevor Pinch (eds.), *Handbook of Science and Technology Studies*, 2005, pp. 652-670.

lent experiments had to face the consequences of his actions in a South-Korean court[7]. But what if the domestic authorities had refrained from dismissing and prosecuting him (after all, government appears to have been involved considerably, although of course not in fabrication and embezzlement, but in funding the "research"[8]) or had been unable to do so (e.g. if all this had taken place in a legally less developed State)? Should unlawful – and even criminal – activities affecting the entire scientific world not be legally reflected also at global level? Would it not be the logical consequence to have such situations governed by international legal standards – and if so, who should formulate and implement them? Moreover: What about the numerous ethical issues and their repercussions in the legal field? Suppose *Hwang Woo-Suk* would really have succeeded in cloning human embryonic stem cells. As is well known, whereas such activity may be legal (and considered to be ethically sound) in that particular Asian country, the legal and ethical situation in other jurisdictions and cultural contexts is an entirely different one. Additionally, during and around the great scandal, the same "scientist" was involved in other ethically doubtful activity, viz. the payment of women donating ova for scientific (?) purposes[9]. Are all these issues outside the scope of action of the international community – are they beyond the reach of international law?

They are not. A closer look reveals the existence of a plethora of international institutions, legal rules and principles, of global norms for the purpose of the international governance of science and of administrative mechanisms to ensure the sound management of science-related problems. We shall discover that neither ethical issues of research, nor the affection of other rights, values and interests by scientific activity, nor the issues related to research funding are ignored by international institutions, international legal norms and global administrative mechanisms. It is these institutions, legal norms and administrative mechanisms we

---

7 David Cyranoski (supra note 4); Park Si-soo, "Hwang Convicted of Embezzlement, Cleared of Fraud", published online 26 October 2006, *The Korea Times*, available at http://www.koreatimes.co.kr/www/news/nation/2009/10/117_54275.html.

8 Péter Kakuk, pp. 553 *et seq.* and 561 (supra note 4).

9 Cf. Robert Steinbrook, "Egg Donation and Human Embryonic Stem-Cell Research", *The New England Journal of Medicine* 354 (4), 26 January 2006, pp. 324-326; Péter Kakuk, p. 547 (supra note 4).

analysed[10] in a research project funded by the German Research Foundation (Deutsche Forschungsgemeinschaft)[11]. This book's purpose is to present the jurisprudential results of the project. Its socio-scientific outcomes have been published separately in German in Sebastian Steinecke's *Zur internationalen Governance der Wissenschaft: Die Regulierung der Wissenschaftsfreiheit zwischen Selbstregelung und hoheitlichem Zugriff – gleichzeitig ein Beitrag zum Wandel von Staatlichkeit*[12]. Empirical results are collected in a free database available at http://www.rewi.uni-jena.de/Fakult_auml_t/Professoren+_amp_+Dozenten/Universit_auml_tsprofessoren/Prof_+Dr_+Matthias+Ruffert/Forschung/Forschungsprofil/Globalisierung+und+Global+Governance/Elemente+eines+transnationalen+Wissenschaftsrechts/Database.html.

The present study is composed of five parts. Firstly, we will give a precise account of the exact field of international legal regulation under scrutiny, which requires substantial effort (below A.). Secondly, we will seize the development of global administrative law and methodologically develop that there is a particular administrative legal field of science (below B.). Thirdly, we will identify freedom of science as the constitutional core of that legal field (below C.). Subsequently, we will comprehensively analyse actors and institutions (below D. and E.). Finally, elements of a global administrative law of science will be summed up and revisited (below F.).

---

10 Together with Katrin Rentzsch and supported by the student assistants mentioned above.

11 *Elemente eines transnationalen Wissenschaftsrechts* (http://gepris.dfg.de/gepris/OCTOPUS/;jsessionid=438B25FE656741D8B11447CB25A494A4?module=gepris&task=showDetail&context=projekt&id=33485187&selectedSubTab=1).

12 München, Herbert Utz Verlag, 2011.

A. The Concept of Science

I. Preliminaries

Analysing the governance of science – whether at the domestic or global level – requires a concept of the term "science"[1]. What appears to be easy at first sight – everyone has at least a vague idea of what science is – proves considerably more difficult once factual and legal connotations of the term are considered in depth. The intrinsic factual particularities of the term set aside temporarily, both main obstacles to its definition in the field of international law are obvious.

Firstly, the notion of science is not used in any particular legal instrument of general recognition. We are well aware that in such universal documents even terminology may be subject to intensive debate, doubtful efforts of definition or continuous uncertainty – considering examples such as "peace" in Article 39 UN Charter[2] or "self-determination" in human rights treaties such as Article 1 ICCPR[3] only. Of course, the notion of science is used in international legal texts (see below E. III.), but there is no single document or even group of documents the quest for a definition can concentrate upon. Consequently, the task of defining the term is part of the effort to design the subject matter itself. There is no positivist approach or else given idea to the concept of international law of science or, more generally, to its legal governance.

Secondly, the term "science" has, without any doubt, different meanings in the various jurisdictions of the world. This is due not only to the quite obvious linguistic divergences, be it between the different languages (science, Wissenschaft, science, sciencia, scienzia, наука, … to name but a few of them) or even within one language (science, scholar-

---

1 Cf. also Sebastian Steinecke, *Zur internationalen Governance der Wissenschaft,* 2010, manuscript pp. 78 *et seq.*

2 Cf. only Jochen Abraham Frowein and Nico Krisch, in: Bruno Simma (ed.), *The Charter of the United Nations. A Commentary, Volume I,* 2nd ed., 2002, Article 39, para. 6.

3 Manfred Nowak, *U.N. Covenant on Civil and Political Rights. CCPR Commentary,* 2nd ed. 2005, Article 1, paras. 32 *et seq.*

ship and academia, Wissenschaft and Wissenschaften...). What is more, is that its importance in law differs from jurisdiction to jurisdiction from mere irrelevance where there are no legal rules governing the field of science to constitutional dignity where there is an explicit guarantee of freedom of science[4]. Discovering the legal framework of governance of science implies the need to find a notion which is at least acceptable to a majority of jurisdictions and also apt to digest the various approaches that may exist in the legal sphere. At any rate, the starting point is outside the law and lies in the historical depth of the term.

II. Science in Context

1. Thought, Philosophy, Method

The idea of science is deeply rooted in the history of the human quest for knowledge, driven by doubt and reflection, aimed at comprehensive understanding of the self and the world around it[5]. In the western world[6], it is supposed to begin with the view of the Presocratics (e.g. *Anaximander, Anaximenes, Heraklit, Pythagoras, Thales*) on nature, methodologically steered by a strong sense of logics and an admiration, if not (continuing) mystification, of numbers and mathematical operations[7]. This found its worthy perpetuation in the abstractive idealism of *Socrates* and *Plato* and the dialectic method so essential for the intellectual reflection, altogether brought to perfection by *Aristotle*[8]. The reception of ancient Greek thinking was constitutive not only for Roman philosophy (above all in the works of *Cicero*), but also for scholastic

---

4 Cf. below C.

5 Comprehensively Sebastian Steinecke (supra note 1), manuscript pp. 71 *et seq.*

6 See above all Lucio Russo, *The Forgotten Revolution,* 2004. For even earlier forms of "science" see Henri Frankfort and Henriette Frankfort, "Myth and Reality", in: id. (eds.), *Before Philosophy. The Intellectual Adventure of Ancient Man*, 1946, pp. 11-36.

7 André Pichot: *Die Geburt der Wissenschaft. Von den Babyloniern zu den frühen Griechen,* 1995, pp. 282 *et seqq*.

8 Cf. Paolo Crivelli, *Aristotle on Truth,* 2006; Jan Szaif, "Die Geschichte des Wahrheitsbegriffs in der klassischen Antike", in: Markus Enders and id. (eds.), *Die Geschichte des philosophischen Begriffs der Wahrheit*, 2006, pp. 1-32.

thought in the Middle Ages (above all *Thomas Aquinas*)[9]. The ages of rationality and enlightenment reinforced the methodological rigidity of calling into question religious, traditional or else given truths, so that the works of *René Descartes* and *Immanuel Kant* can particularly be considered the methodological core of western thinking[10].

It is in this epoch at the latest that the history of scientific thinking is interwoven with the history of universities as a particular institution for the promotion and proliferation of science[11]. The achievements of the brothers *Humboldt* are crucial not only from a germanocentric perspective (which shall of course be avoided), but in the reforms triggered in particular by *Wilhelm von Humboldt* we can see a culmination of both the idea of the university up to his time (with places such as Bologna, Paris, Oxford and Cambridge as predecessors outside Germany to be mentioned by all means) and the world-wide success of that very idea in modern times: it is well known that newly founded universities in the United States of America (Johns Hopkins University, Baltimore 1876; University of Chicago, 1890; California Institute of Technology, 1891) took up the *Humboldtian* ideal and that traditional American institutions of higher education (such as Harvard and Princeton) shifted towards this ideal (and away from British and French examples) after the downfall of the Napoleonic empire[12].

---

9 Thomas Aquinas, *The Disputed Questions On Truth,* Vol. I, translated by Robert William Mulligan, 1952.

10 On Descartes cf. Ferdinand Alquié, *Wissenschaft und Metaphysik bei Descartes,* 2001; Hans Radermacher, *Cartesianische Wissenschaftstheorie,* 1971. For Kant cf. only Immanuel Kant, "The Contest of Faculties", in: Hans Siegbert Reiss (ed.), *Kant: Political Writings*, 2nd ed. 1991, pp. 176-190.

11 Cf. Helmut Schelsky, *Einsamkeit und Freiheit. Idee und Gestalt der deutschen Universität und ihrer Reformen,* 1963.

12 Ronald Standler, *Academic Freedom in the USA*, 1999, available at www.rbs2.com/afree.htm; Hermann Röhrs, *Der Einfluss der klassischen deutschen Universitätsidee auf die Higher Education in Amerika*, 1995, pp. 73-85; Roy Turner, "Humboldt in North America? Reflections on the Research University and its Historians", in: Christoph Schwinges, *Humboldt International. Der Export des deutschen Universitätsmodells im 19. und 20. Jahrhundert*, 2001, pp. 289-312 at pp. 289 *et seq*.; Walter Metzger, "The German Contribution To The American Theory Of Academic Freedom", *American Association of University Professors Bulletin* 41 (1955), pp. 214-230, printed in: id. (ed.), *The American Concept of Academic Freedom in Formation. A Collection of Essays and Reports*, 1977; Balakrishnan Rajagopal, "Academic Freedom as a Human

Given this universal reach of the *Humboldtian* idea of the university and of science, it is justified to take up some of its content in defining what science means as an object of governance and legal regulation. According to his famous dicta, solitariness and freedom lay the foundations of scientific thought – the independent, reflective and free activity of the single thinker[13]. Science – *Wissenschaft* – in this tradition is the never ending, serious and methodologically planned quest for truth, as the German Federal Constitutional Court (*Bundesverfassungsgericht*) defined with reference to *Humboldt* and the interpretation of his works by the early 20th century legal scholar *Rudolf Smend*[14]. Again, we submit that these ideas are not intrinsic to German philosophic and legal thought but are designed to convey a universal concept of what is encompassed in science[15].

Along the same lines as *Humboldt*, *Robert Merton* in his work on the sociology of science, undertakes to define science by means of four elements (often known as "CUDOS" for the first letters of the respective terms): (1) communalism – all scientifically gained knowledge has to be accessible for free debate and scientists renounce intellectual property rights in exchange for reputation, (2) universalism – the quality of science to stand intersubjective control, (3) disinterestedness – the absence of any pecuniary or otherwise material interest and finally (4) organised scepticism – all scientific results have to be able to be called into question at all times[16]. *Merton's* concept has been criticised to be

---

Right. An Internationalist Perspective", in: *Academe* Vol. 89, issue 3 (May-June 2003), pp. 25-28 at p. 26.

13 Cf. Helmut Schelsky (supra note 11).

14 *Entscheidungen des Bundesverfassungsgerichts* 35, 79 at p. 113, recurring upon Wilhelm von Humboldt, *Über die innere und äußere Organisation der höheren wissenschaftlichen Anstalten in Berlin* (1809/10), quoted in: Ernst Anrich (ed.), *Die Idee der deutschen Universität,* 1956, pp. 375-386 at p. 379: science as "... etwas noch nicht ganz Gefundenes und nie ganz Aufzufindendes" – *something not yet found and never really to be found*, in the interpretation by Rudolf Smend, "Das Recht der freien Meinungsäußerung", *Veröffentlichungen der Vereinigung der Deutschen Staatsrechtslehrer* 4 (1928), pp. 44-74, at p. 67.

15 The requirement of "methodological plannedness" does not exclude random results (which took place in the history of science, e.g. in the discovery of penicillin and X-ray, cf. Sebastian Steinecke (supra note 1), manuscript pp. 82 and 84.

16 Robert K. Merton, "Science and Technology in a Democratic Order", *Journal of Legal and Political Sociology* 1 (1942), pp. 115-126.

too idealist[17], but this should not exclude it from our perspective since it has gained large influence and is at least partly subject to general consent. Also, the current approach of the American Physical Society is *Humboldtian* in its methodological perspective, as it defines:

> "Science is the systematic enterprise of gathering knowledge about the universe and organizing and condensing that knowledge into testable laws and theories."[18]

Of course there is further development of methodology and content of scientific thought after *Humboldt.* It may be fair to say that such categories as the theory of science or sociology of science did not come into being before very recently[19]. But instead of going into the seminal works of authors like *Gottlob Frege, Alfred Tarski, Charles Sanders Pierce, Jürgen Habermas* or *Wilhelm Kamlah*[20], an important point for legal analysis has to be made: While it is true without any doubt that science is about serious human reflection on certain problems, it does not include all such reflection. To take up a *bon mot often used,* which criticises the *Humboldtian* approach: A police officer investigating into a crime tries to acquire knowledge by methodologically sound and serious reflection – but he is certainly not a scientist[21]. Beyond such obvious exclusions, other reflective activity has to be set aside, such as political debate or literary thought, which may also be subject to governance and legal regulation, but in other fields of the law, with different purposes and distinct legal limits. Intellectual reflection, philosophical thought, intellectual exchange of ideas may take place in scientific con-

---

17 Cf. S. Barry Barnes and R.G.A. Dolby, "The Scientific Ethos: A Deviant Viewpoint", *Archives Européennes de Sociologie* XI (1970), pp. 3-25; Harriet Zuckerman, "Sociology of Science", in: Neil Joseph Smelser (ed.), *Handbook of Sociology,* 1988, pp. 511-574 at pp. 517 *et seq.* A reason for the strong ethical orientation of Merton's approach is its direction against the oppression of academic freedom in totalitarian regimes (Peter Weingart, *Wissenschaftssoziologie,* 2003, at pp. 15 *et seq.*, in particular at p. 19).

18 American Physical Society, *Statements on Ethics and Values,* Nr. 99.6 "What is Science?" (1999).

19 Cf. only Alan Francis Chalmers, *Science and its Fabrication,* 1990; id., *What is this thing Called Sciences,* 3rd ed. 1999.

20 Cf. Sebastian Steinecke (supra note 1), manuscript pp. 154 *et seq.*

21 Hans Joachim Schneider, *Kriminologie für das 21. Jahrhundert – Schwerpunkt und Fortschritte der internationalen Kriminologie*, 2001, p. 115; Christian Starck, in: Hermann von Mangoldt/Friedrich Klein/id. (eds.), *Kommentar zum Grundgesetz*, Vol. 1, Article 5, para. 352.

texts, but this is not necessarily so. Nonetheless, what can be stated as a result of this *tour d'horizon* through western intellectual history is that science as a potential object for legal governance is about the methodologically sound creation of knowledge with a general purpose.

2. Technology

Most institutions, principles and rules to be analysed in this book can be related to such an idealistic notion of science only with great difficulties. Undoubtedly, scientific research in modern times is to a vast extent linked to the creation of technological development – and to its economic benefits, be it of researchers, commercial applicants or users of scientifically gained products. Applied technological science does not quest for truth, but is designing reality[22]. It is as much part of the self-image of the scientific world as of the perception of society at large that visible effects of research in technological and finally economic terms are part of the matter. Any concept of the law of science excluding such applied research would be imperfect, if not outside social reality[23]. Efforts in research have largely shifted from universities or public entities (such as the noble *Academies* of former times[24]) towards private business[25], whether as such or in particular forms of public-private-partner-

---

[22] Matthias Ruffert, "Grund und Grenzen der Wissenschaftsfreiheit", *Veröffentlichungen der Vereinigung der Deutschen Staatsrechtslehrer* 65 (2006), pp. 146-210 at p. 157 (recurring upon Ralf Kleindiek, *Wissenschaft und Freiheit in der Risikogesellschaft,* 1998, pp. 128 *et seq.;* following Hans-Peter Dürr, *Das Netz des Physikers,* 1988, pp. 10 *et seq.*). Cf. also Karin Knorr-Cetina, *The Manufacture of Science,* 1981.

[23] Nonetheless, such concepts are proposed by David Lindberg, *Die Anfänge des abendländischen Wissens,* 2000, pp. 1 and 6, and Klaus Pähler, *Qualitätsmerkmale wissenschaftlicher Theorien,* 1986, p. 2.

[24] Cf. Marta Ornstein, *The Role of Scientific Societies in the Seventeenth Century,* 1975.

[25] The bulk of expenses in research efforts is spent here: Helmuth Schulze-Fielitz, "Politische Voraussetzungen wissenschaftlicher Forschung", in: Horst Dreier and Dietmar Willoweit (eds.), *Wissenschaft und Politik,* 2010, pp. 71-106 at p. 77 *et seq.*

ships[26]. It can be shown that a considerable part of international legal regulation in science is mainly applicable to such result-driven research on an economic background. To give but one prominent example: One of the few comprehensive texts on the international governance of science, the (not legally binding) "Frascati Manual" of the OECD defines researchers as[27]

> "... professionals engaged in the conception or creation of new knowledge, products, processes, methods, and systems, and in the management of the projects concerned."

The proximity of such notion of the researcher to science in a technological and economic context is obvious, considering both the institution issuing that definition and the content of the definition. It should finally be added that there is no reason at all to "downgrade" such research in legal or even in moral terms, given that the entanglement of scientific and economic activity can be proven even historically[28].

3. Scholarship?

At this point at the latest, the linguistic trap has to be efficiently avoided. Readers from the Anglo-Saxon world could easily criticise the approach of this book towards science for lack of precision, looking with less criticism to what has been said on science, technology and economy, but with more harsh reproaches against the inclusion of fields such as philosophy, history – or jurisprudence. It has already been mentioned that there are gaps and even trenches between "science" (in English but also in French) on the one hand and notions such as "Wissenschaft" and "наука" on the other hand, the former being confined to research activity related to nature and technology, the latter referring also to what would be called "scholarship" within the "humanities" in Eng-

---

26 Cf. Ulrich Hilpert, "The State, Science and Techno-Industrial Innovation. A New Model of State Policy and a Changing Role of the State", in: id (ed.), *State Policies and Techno-Industrial Revolution,* 1991, pp. 3-40 at pp. 10 *et seq.*

27 OECD, *The Measurement of Scientific and Technological Activities, Proposed Standard Practice for Surveys of Research and Experimental Development, Frascati Manual,* 1993, p. 86.

28 See already Friedrich Schiller (together with Johann Wolfgang von Goethe), "Xenien aus dem Musen-Almanach für das Jahr 1797", in: Friedrich Schiller, *Gedichte* (edited by Georg Kurscheidt), 1992, pp. 577-629, at p. 585.

lish[29]. But if we were to further elaborate on this distinction, it proves flawed already in linguistic terms: a professor of philosophy or law would perhaps not be considered a "researcher" in the English speaking world, but certainly a "chercheur" in France (though there is no "science de la philosophie" or "du droit" in French). After all, the international governance of the field of research has to be open to different jurisdictions and their linguistic approaches. This book will therefore not be unaware of terminological divergences and diversities, but it will also not take them as the basis of exclusive operations. On the contrary: The effects produced by different understandings of "science" will be shown in parts of the book.

III. Scientific Revolutions and the Scientific Community

Science is not only an individual activity, but a social phenomenon[30]. The recognition of an activity as scientific research by the community of researchers, the scientific community, is crucial for the description of what is science. This aspect proves helpful to exclude many activities of the quality of everyday reflection (the above-mentioned police investigation) and also of "pseudo-science". For ages, wise men have tried to produce gold, to predict individual and collective faith from the position of celestial bodies or to heal diseases by applying magnetic forces. Neither alchemy nor astrology nor mesmerism are considered to be sciences, though, for the very reason that they lack recognition by the scientific community for obvious reasons[31]. A similar approach may be

---

[29] On this tradition cf. Wissenschaftsrat, *Empfehlungen zur deutschen Wissenschaftspolitik im Europäischen Forschungsraum*, 2010 (Drucksache 9866-10), at p. 20.

[30] Cf. Helga Nowotny, "The Changing Nature of Public Science", in: id./Dominique Pestre/Eberhard Schmidt-Aßmann/Helmuth Schulze-Fielitz/Hans-Heinrich Trute, *The Public Nature of Science under Assault*, 2005, pp. 1-27. Cf. also Rudolf Stichweh, "The Multiple Publics of Science: Inclusion and Popularization", *Soziale Systeme* 9 (2003), pp. 210-220; on the popularization of science.

[31] Cf. on astrology Bart Bok and Lawrence Jerome (eds.), *Objections to Astrology*, 1975; Paul Thagard, "Why astrology is Pseudoscience", *Proceedings of the Biennial Meeting of the Philosophy of Science Association* 1 (1978), p. 223-234. This does not exclude that these "sciences" produced results that could be used in the recognised natural sciences, cf. William Newman, *Atoms and Al-*

taken towards intellectual constructions taking some inherent "truths" for granted either for religious reasons (e.g. creationism) or due to fixed ideological orientations (e.g. certain Marxist tendencies or those who deny the existence of the Shoah).

But implying the perspective of the international communities is not devoid of risk. What if the contemporary scientific communities of *Nikolaus Kopernikus* and *Galileo Galilei* had been asked to assess the research activity of their colleagues? What if *Isaac Newton* and his contemporaries had had the opportunity to subdue *Albert Einstein's* theories under a similar assessment? In his seminal work on scientific revolutions, *Thomas Kuhn* shows that research may be undertaken in two ways: (1) "standard science" following a certain scientific paradigm and (2) research leading to a change of paradigm, thus to a scientific revolution[32]. Breaking new ground and overturning hitherto recognised buildings of knowledge is an integral part of the most important research activities and their results. Thus, recognition and acceptance within the scientific community must not be given overall and absolute importance, but may themselves be called into question[33].

For the purpose of international legal governance, it is rarely necessary to draw a distinct line between science and "pseudo-science", though. In most instances, the reference to the perspective of the scientific community will be a reliable indicator. If, however, this perspective leads to the exclusion of a person or activity from the field of science, and if this implies legal consequences, the perception of the scientific community cannot be taken for granted without closer scrutiny.

IV. Science and the Law

Whatever the role of the scientific community, science has its own rules. Not only is it impossible to predict where the quest for new knowledge leads the scientist and the general public, but it is also impossible to regulate the scientific process as such. The law can create space for free scientific research, it can erect institutions that promote scientific activ-

---

chemy. Chemistry and the Experimental Origins of Scientific Revolution, 2006, and Alison Winter, *Mesmerized. Powers of Mind in Victorian Britain*, 1998.

32 Thomas Kuhn, *The Structure of Scientific Revolutions*, 3rd ed. 1996, p. 23.

33 Cf. Roy Wallis (ed.), *On the Margins of Science: The Social Construction of Received Knowledge*, 1979.

ity and it can set the legal framework to provide material resources for science – but it cannot order scientific progress to take place[34]. Moreover, legal restrictions may interfere with the free creation of the results of scientific research. In a way, the autonomy of the scientific world is vested with a degree of hostility towards legal regulation[35]. In States with a democratic constitution and in an international legal sphere which considers the democratic creation of law as a value as such, this juxtaposition between law and science can be described as a provoking tension between science and democracy[36].

V. A Tentative Definition

All in all, this book analyses the international legal governance of science which is considered as the reflected, autonomous quest for new knowledge that can be integrated into existing systems of knowledge or bears the capacity to overcome them, notwithstanding the technological or economic applicability of such knowledge[37].

---

[34] For a recent assessment cf. Helmuth Schulze-Fielitz, "Politische Voraussetzungen wissenschaftlicher Forschung", in: Horst Dreier and Dietmar Willoweit (eds.), *Wissenschaft und Politik,* 2010, pp. 71-106.

[35] Matthias Ruffert (supra note 22), p. 160 *et seq.*

[36] Matthias Ruffert (supra note 22), at p. 161. This is in a certain contradiction with the assumption that a more democratic a society, the more there is free science. This assumption is propounded by international institutions such as the Council of Europe (CM/AS(2007)Rec1762 final of 1 October 2007: "The Committee of Ministers believes that academic freedom and university autonomy are among the indicators which measure how democratic a society is.") as well as by academic writers (Robert Merton, "The Normative Structure of Science", in: Norman Storer (ed.), *The Sociology of Science: Theoretical and Empirical Investigations,* 1973, pp. 267-280 at p. 269; David Hollinger, "The Defense of Democracy and Robert K. Merton's Formulation of the Scientific Ethos", *Knowledge and Society: Studies in the Sociology of Culture Past and Present* 4 (1983), pp. 1-15, as well as Ronald Tobey, *The American Ideology of National Science 1919–1930,* 1971, Chapter 2 *et seq.*; André Pichot (supra note 7), pp. 547 *et seq.*, who tries to prove the assertion mentioning the fact that science first blossomed in democratic ancient Greece). Also democratically created legislation is able to restrict science if it voluntarily (or even accidentally) interferes with its inherent autonomy.

[37] Sebastian Steinecke (supra note 1), manuscript p. 182.

B. Global Administrative Law

I. An Emerging Concept for the Legal Analysis of the Governance of Science

The idea of developing administrative legal structures at a global level is a prominent conceptual invention in current public law thinking. The theoretical approach to global administrative law aims at the elaboration of interrelationships between administrative entities on a world scale, the identification and analysis of decision-making processes and law-creating mechanisms in the divergent systems of legal sources, be they international or domestic, and their scrutiny according to the different tasks which can be ascribed to an administrative system[1]. At the heart of the concept is the legally sound accomplishment of different policy tasks at global level by mechanisms that can be construed more or less analogously to those mechanisms that are pertinent in domestic administrative (legal) systems[2]. *Sabino Cassese*, one of the most influential proponents of the idea of global administrative law, goes so far even as to draw a parallel between the current debate and the emergence of administrative law as a domestic legal concept by authors such as *Edouard de Laferrière* and *Otto Mayer*[3]. At the same time, the related

---

1 Nico Krisch, "The Pluralism of Global Administrative Law", *European Journal of International Law* 17 (2006), pp. 247 *et seqq.*; Benedict Kingsbury/Nico Krisch/Richard B. Stewart, "The Emergence of Global Administrative Law", *Law and Contemporary Problems* 68 (2005), pp. 15 *et seqq.* at p. 17. See also Daniel C. Esty, "Good Governance at the Supranational Scale: Globalizing Administrative Law", *Yale Law Journal* 115 (2006), pp. 1490-1562.

2 Sabino Cassese, "Administrative Law without the State? The Challenge of Global Regulation", *New York University Journal of International Law and Politics* 37 (2005), pp. 663-694. at pp. 668 *et seq.*

3 Sabino Cassese, "Is There a Global Administrative Law?", in: Armin von Bogdandy/Rüdiger Wolfrum/Jochen von Bernstorff/Philipp Dann/Matthias Goldmann (eds.), *The Exercise of Public Authority by International Institutions*, 2010, pp. 761-776.

concept of international administrative law sees an impressive renaissance in recent legal scholarship[4].

The theory of global administrative law focuses on the plurality of administrative regimes with their plethora of actors – public or private bodies[5], their diffuse bulk of legal rules – binding and non-binding – and the concomitant issues of the protection of rights and interests as well as of legitimacy. According to the concept of global administrative law such plurality needs a rational and legally sound sharing of responsibilities within the different global regulatory systems[6].

Such a concept is the ideal analytical framework for the international governance of science. As we shall see in the later chapters, the empirical situation of the governance of science corresponds exactly to what global administrative lawyers focus at: A great variety of (public and private) actors emits an even greater variety of (binding and non-binding) rules which is then implemented by the same actors by means of administrative mechanisms (compulsory and non-compulsory). We follow the approach that without the identification of sound public law requirements for the protection of individual rights and the assurance of legitimate exercise of unilateral government, central values of the concept of public law would be at peril[7], and that the elaboration of global administrative law can lead to such protection and guarantee.

---

[4] Cf. the comprehensive studies by Claus-Dieter Classen and Giovanni Biaggini, "Die Entwicklung eines Internationalen Verwaltungsrechts als Aufgabe der Rechtswissenschaft", *Veröffentlichungen der Vereinigung der Deutschen Staatsrechtslehrer* 67 (2008), pp. 365-412 and 413-445.

[5] Cf. the typology concerning organisation in Markus A. Glaser, *Internationale Verwaltungsbeziehungen,* 2010, at pp. 28 *et seq.* See also Armin von Bogdandy/Philipp Dann/Matthias Goldmann, "Developing the Publicness of Public International Law: Towards a Legal Framework for Global Governance Activities", in: Armin von Bogdandy et al. (supra note 3), pp. 3-32 at pp. 13 *et seq.*, on the importance to include private bodies.

[6] Nico Krisch (supra note 1), at pp. 269 *et seq.*

[7] Armin von Bogdandy/Philipp Dann/Matthias Goldmann (supra note 5), passim.

II. Conceptualisation

1. Public International Law and Global Administrative Law

A new concept has to clarify its relationship to existing ones, whether it is interlinked to or separated from them. The conceptualisation of global administrative law is bound to fail if it remains unclear where its place in legal scholarship should be. This is a question not only of terminology (below 2.), but also of methodology, argumentation and even style. In this respect, formulating the place of global administrative law within public international law is of utmost importance.

It is a platitude, of course, that the concept of international law has profoundly changed within the past few years and decades. We are facing a shift in the legal substance, but above all in scholarly methods to handle that substance, caused partly by new approaches to international relations theory and other fields of social sciences and also political philosophy. It has become common to designate three steps in this process: from (1) co-ordination to (2) co-operation to (3) community as a *leitmotiv* of international law[8]. Nonetheless, this is not a scale on which consecutive steps replace former developments. The traditional power-related view is by no means excluded from the perspective. It appears to be generally accepted – and rightly so – that different layers of international law, originating from different epochs, are co-existing side by side[9]. It is therefore methodologically sound not to give up realist views in power-oriented fields with normative contents rooted in the traditional, "Westphalian" system whilst bringing forward idealist and institutionalist perspectives in areas of greater value-orientation or institutional density[10].

---

[8] Classical work: Wolfgang Friedmann, *The Changing Structure of International Law,* 1964. Cf. also Anne Peters, "Global Constitutionalism in a Nutshell", in: Klaus Dicke/Stephan Hobe/Karl-Ulrich Meyn/id./Eibe Riedel/Hans-Joachim Schütz/Christian Tietje (eds.), *Weltinnenrecht,* 2005, pp. 536 *et seqq.*, and the very illustrative description by Tim Wihl, "Freiheit als Unwert? Verwandlungen des Völkerrechts aus liberaler Perspektive", in: Christian Tomuschat (ed.), *Weltordnungsmodelle für das 21. Jahrhundert,* 2009, pp. 65-98 at p. 72.

[9] Antonio Cassese, *International Law,* 2nd ed., 2005, at p. 21.

[10] In the context of global administrative law: José E. Alvarez, *International Organisations as Law-makers,* 2005, at pp. 244 *et seq.*

What global administrative law aims at in this respect is the further elaboration of the third, communitarian layer of public international law in continuing the establishment of its capacity to legitimate and to limit the exercise of power on a global scale[11]. Global administrative law proposes to effectuate this by extending basic public law functions to the international sphere.

If we further follow this path, applying an international legal pattern to science as understood here becomes viable. In "classic" – co-ordination and even co-operative – terms of Public International Law, there is no necessity to consider a field of reality such as science and to analyse its particular legal framework[12]. International law is about States (or else other subjects endowed with legal personality), their external powers and the interrelationship of such powers. A legal perspective on the international governance of science would only be possible under these assumptions, if there was a comprehensive international treaty or an all-encompassing international organisation, but, as we know, such simple legal or institutional framework does not exist. The perspective of global administrative law, on the contrary, offers a conceptual framework for the legal analysis of the global governance of science.

2. Terminology: International, Transnational, Global

The term "global administrative law" is not devoid of ambivalence, though[13]. A first terminological uncertainty arises with respect to the epithets "international" and "global". Is "global administrative law" distinct from "international administrative law", or are we faced with the same phenomenon under different titles? Two reflections will provide at least some clarification.

The first reflection refers to the use of the term "international administrative law" in public international legal scholarship. In a traditional legal context, the activity of international organisations has been analysed

---

[11] Armin von Bogdandy/Philipp Dann/Matthias Goldmann (supra note 5) at p. 10.

[12] Consider, for the change in contents of Public International Law, José E. Alvarez (supra note 10) at pp. 601 *et seq.*

[13] For an overview: Franz Mayer, "Internationalisierung des Verwaltungsrechts?", in: Christoph Möllers/Andreas Voßkuhle/Christian Walter (eds.), *Internationales Verwaltungsrecht*, 2007, pp. 49-72 at pp. 54 *et seq.*

from an administrative law perspective, focussing upon internal matters of such organisations, e.g. legal relationships towards the organisations' personnel, budgetary affairs or internal dispute resolution (administrative tribunals of the UN or other organisations[14])[15]. A more topical perspective concentrates upon administrative relationships created in the external activity of international organisations together with administrative structures that do not amount to the creation of such organisations[16]. Networks of authorities without involvement of the State as an overarching entity are of particular concern to this approach[17], and the proximity to the idea of transnationality (see below) is obvious. The international regulation of financial markets (Basle Committee), framework rules on interregional transboundary co-operation or even co-operation with private actors (e.g. multi-national companies) are just examples of how the administrative law approach can be practically applied[18]. It is about the foundation and limitation of power on the global scale, and it is submitted that the difference between international and domestic institutions is not crucial for the application of public law rules as a matter of principle[19]. To sum up the first reflection, little can be pleaded to separate global from international administrative law[20].

The second reflection focuses upon a parochial terminological (and doctrinal) development. In German legal writing, the terminology has for a long time been blocked by a particular use of the term "international administrative law" (*Internationales Verwaltungsrecht*), being

---

[14] On these cf. Benedict Kingsbury/Nico Krisch/Richard B. Stewart (supra note 1) at p. 20 (footnote 11).

[15] In this context Chittharanjan Felix Amerasinghe, "The Future of International Administrative Law", *International and Comparative Law Quarterly* 45 (1996), pp. 773 *et seqq.*; Christine Breining-Kaufmann, "Internationales Verwaltungsrecht", *Zeitschrift für Schweizerisches Recht* 125 (2006), II, pp. 5-73 at p. 10.

[16] Cf. Christian Walter and Matthias Ruffert, *Institutionalisiertes Völkerrecht,* 2009, para. 661.

[17] Akin to that is the approach of Christoph Möllers, "Transnationale Behördenkooperation", *Zeitschrift für ausländisches öffentliches Recht und Völkerrecht* 65 (2005), pp. 351-389.

[18] Seminal work: Christian Tietje, *Internationalisiertes Verwaltungshandeln,* 2001.

[19] This is feared by Armin von Bogdandy/Philipp Dann/Matthias Goldmann (supra note 5) at pp. 24 *et seq.*

[20] Another view is taken by Giovanni Biaggini (supra note 4) at p. 419.

understood above all as a parallel to "private international law" (i.e. conflict of laws in private law) and thus related to rules of collision between jurisdictions and the applicability of the law of a certain country to a given case[21]. Unlike in other jurisdictions, these questions have for a long time continued to be highly controversial: Is it really possible to ask the classical question about the applicability of foreign law in a field which is completely dominated by the activity of national authorities which are above all bound by the requirements of the respective national public law?[22] Although that controversy has never been completely overcome, it has certainly become less important by now[23]. German scholarship tends to shift the focus towards a tripartite concept created by *Eberhard Schmidt-Aßmann* differentiating between law (1) of international administrative institutions, (2) determinative of national administrative legal orders and (3) cooperative handling on multilevel issues[24]. It is by no means impossible to integrate conflict-of-laws questions into such concepts whenever they should arise. To give an example from the field of science: The execution of a bio-ethically doubtful research project by a multinational research institution could be governed either by the bio-ethical rules of an international organisation or by conflicting rules of different nation States (the State where the institution is seated, where the project is mainly performed, where the researchers originate from...). Before searching a solution to such issues, addressing them comprehensively is obviously advantageous. This, in turn, reduces the risk of international administrative law to be misun-

---

21 The development is explained and illustrated by Christian Tietje (supra note 18), para. 662. The leading work is still Karl Neumeyer, *Internationales Verwaltungsrecht, Vol. IV, Allgemeiner Teil,* 1936, at pp. 105 *et seq.*,115 *et seq.* and 121 *et seq.* The approach is recently focused upon by Dirk Ehlers, in: Hans-Uwe Erichsen and id. (ed.), *Allgemeines Verwaltungsrecht*, 14th edition 2010, § 4.

22 The brilliant critique by Klaus Vogel, *Der räumliche Anwendungsbereich der Verwaltungsrechtsnorm,* 1965, p. 298 *et seq.* has to be mentioned.

23 A reconciliatory position is taken by Christoph Ohler, *Die Kollisionsordnung des Allgemeinen Verwaltungsrechts,* 2005, p. 3; and Matthias Ruffert, "Perspektiven des Internationalen Verwaltungsrechts", in: Christoph Möllers/Andreas Voßkuhle/Christian Walter (eds.) (supra note 13), pp. 395-420.

24 *Eberhard Schmidt-Aßmann*, "The Internationalization of Administrative Relations as a Challenge for Administrative Law Scholarship", in: Armin von Bogdandy et al. (eds.) (supra note 3), pp. 943-964; original German version: "Die Herausforderung der Verwaltungsrechtswissenschaft durch die Internationalisierung der Verwaltungsbeziehungen", *Der Staat* 45 (2006), pp. 315-338.

derstood as being related to issues of jurisdictional conflict only. To sum up this second reflection, the term "international administrative law" is not misleading in such a way that it could not be used besides "global administrative law", and to conclude the "international-global-issue", we submit that the notions are practically synonymous and viably interchangeable.

Left open is the notion of transnationality, which by now is older than 50 years already. Ever since 1956, when *Philip C. Jessup* used the term "transnational law" to point out that the international legal relations could not comprehensively be understood by considering only those relationships between States (nations), i.e. international[25], it has been obvious that sub-state entities and private actors must somehow be integrated in the legal sphere beyond domestic jurisdictions. This perspective particularly focuses on relationships transcending state borders and limits of jurisdictions, and it is by no means necessary to exclude traditional international legal relationships. Therefore, transnationality is concomitant with a holistic view of the non-parochial legal world[26].

The idea of transnationality, the integration of private actors into the international legal field is flawed only in one particular instance. It bears the risk of downgrading essential differences between public (the State, sub-State public bodies) and private (companies, associations) legal subjects. The fundamental distinction between liberty on the one hand, borne by individuals including their private corporate emanations, and authority on the other hand, vested in public institutions that have to be legitimised, should by all means be upheld by any approach towards the international legal sphere[27].

If this particular risk is not overlooked, the idea of transnationality is extremely useful for the analysis of the legal governance of science. States are actors in the field, and State operated or at least funded research is not exceptional. Nevertheless, leaving aside a concept which is explicitly designed to integrate private actors and legal relations be-

---

[25] Philip C. Jessup, *Transnational Law*, 1956. Cf. in particular Christian Tietje/Alan Brouder/Karsten Nowrot, *Philip C. Jessup's Transnational Law Revisited – On the Occasion of the 50th Anniversary of its Publication*, 2006.

[26] The concept of transnationality is taken up in an administrative law context by Andreas Fischer-Lescano, "Transnationales Verwaltungsrecht", *Juristenzeitung* 2008, pp. 373-383.

[27] Cf. Armin von Bogdandy/Philipp Dann/Matthias Goldmann (supra note 5).

tween subjects other than States would lead to a failure in considering the great variety of institutions, public or private, which are engaged in research activity. On the contrary: the great majority of non-domestic legal relationships concerning science are transnational in *Jessup's* sense. Transboundary relationships between universities, research funding organisations and with or between private companies acting in the field of research transcend jurisdictional limits and are not dependent upon State activity as such. Although *Jessup* intended to give a distinct meaning to the term "transnational", it is often used as a synonym for "international" in a broader sense now – which is due to its success: it is a commonplace that transnational relationships have to be integrated into a sound and viable modern concept of international law – and also into global administrative law.

All in all, terminological divergences must not be exaggerated. Global administrative law is an heir to many productive predecessors, be it the shift in international legal scholarship, the analysis of conflict of laws in public law or the discovery of transnational legal relationships. If the term *global* administrative law is used, this is done by virtue of practicality and synthesis rather than exclusion. Consequently, should the terms "international" or "transnational" be used occasionally, this is not meant to deviate from the overall approach, but to emphasise different aspects of one and the same development.

3. Global Administrative Law and Global Governance

Another approach the advantages of which cannot easily be discarded is the more recent concept of global governance that has entered international legal thinking from different directions of the social sciences. Though it is indispensable in the analysis of the management of scientific activity on the global scale, its particular function has to be borne in mind.

The concept of governance is important due to its analytical, descriptive content mainly. Governance refers to situations and processes of governing in complex structures[28]. Using the idea of governance, institu-

---

[28] Seminal works: Jan Kooiman (ed.), *Modern Governance: New Government-Society Interactions,* 1993; James N. Rosenau and Ernst Otto Czempiel (eds.), *Governance without Government,* 1992. For comprehensive overviews see Gunnar Folke Schuppert and Michael Zürn (eds.), "Governance in einer

tions and instruments of hierarchical government, network structures and mechanisms of self-regulation can be analysed comprehensively[29]. The key change of perspective is from actors to regulatory structures[30]. For the sake of completeness, it has to be added that the normative content of the term governance as it appears in the "good governance" policy of the World Bank is of minor importance here[31].

By the term *global governance,* the concept of governance is transferred to the level beyond the domestic sphere, and it is firmly established in the social sciences, albeit definitions offered by prominent authors (such as: "Governance is the sum of the many ways individuals and institutions, public and private, manage their common affairs."[32]) are still somehow vague. When reflections on global governance are undertaken in international legal literature, regulative structures are usually analysed in certain fields of global importance such as economy, environment, security or communication[33]. What is important is the development of problem-solving capacity in an organised plurality of actors[34]. It is therefore astonishing – to add a marginal note – that the global

---

sich wandelnden Welt", *Politische Vierteljahresschrift-Sonderheft* 41/2008; Gunnar Folke Schuppert (ed.), *Governance als Prozess,* 2008.

[29] Christian Walter and Matthias Ruffert, *Institutionalisiertes Völkerrecht,* 2009, para. 657.

[30] Hans-Heinrich Trute/Doris Kühlers/Arne Pilniok, "Der Governance-Ansatz als verwaltungsrechtswissenschaftliches Analysekonzept", in: Gunnar Folke Schuppert and Michael Zürn (eds.) (supra note 28), pp. 173-189 at p. 174, as explained by Gunnar Folke Schuppert, "Was ist und wozu Governance", *Die Verwaltung* 40 (2007), pp. 461-511 at pp. 483 *et seq.*

[31] Sabine Schlemmer-Schulte, "Internationales Währungs- und Finanzrecht", in: Christian Tietje (ed.), *Internationales Wirtschaftsrecht,* 2009, paras. 9/55 *et seq.* Christian Theobald, "Die Weltbank: Good Governance und die Neue Institutionenökonomik", *Verwaltungsarchiv* 89 (1999), pp. 467-487.

[32] James Rosenau, *Along the Domestic-Foreign Frontier – Exploring Governance in a Turbulent World,* 1997, p. 10 *et seq.* Cf. also the Commission on Global Governance, *Our Global Neighbourhood,* 1995, p. 2.

[33] Cf. the studies in Christoph Möllers/Andreas Voßkuhle/Christian Walter (eds.), *Internationales Verwaltungsrecht,* 2007.

[34] Matthias Ruffert, *Die Globalisierung als Herausforderung an das Öffentliche Recht,* 2004, pp. 31 *et seq.* On the reception of the idea of governance in international law see Joseph H. H. Weiler, "The Geology of Internatonal Law – Governance, Democracy and Legitimacy", *Zeitschrift für ausländisches öffentliches Recht und Völkerrecht* 64 (2004), pp. 547-562 at p. 559 *et seq.*

governance of science has not been analysed until now, whether in the social sciences or in international law[35]. The state-of-the-art still appears to be represented by the famous 1992 article of the former President and Judge of the ICJ *Manfred Lachs* about "Thoughts on Science, Technology and World Law"[36] – valuable thoughts, scholarly well-founded thoughts, but (taking up the modesty of their author – who died shortly after their publication) thoughts only.

There continues to be unease in legal literature concerning the viability of the concept to be used as an instrument of analysis in public international law. Whereas the analytical advantages of the idea of governance for the description of reality are accepted, scholars underline its deficiencies in formulating clear legal requirements for the legitimation and limitation of the exercise of power[37]. It should however be pointed out that this does not really undermine the analytical value of global governance. It is literally impossible to find a way through the intricate fabric woven by the plethora of actors and the variety of principles of rules in the international field of science without an analytical perspective that goes beyond a State-centred approach. There is no need to set aside the concentration on regulatory structures brought forward by governance theorists to preserve the rights and values under protection by legal principles if only the legal requirements are not omitted as a next step. In this regard, the governance approach is used as a layer of reality beneath the legal safeguards to be implemented by means of global administrative law.

4. Global Administrative Law and Global (Multilevel) Constitutionalism

It might be argued that these principles, rights and values to be implemented through the elaboration of global administrative law are in their essence constitutional ones, and of course the idea of global constitu-

---

35 Consider the comprehensive book of Volker Rittberger and Bernhard Zangl, *Internationale Organisationen*, 3rd ed. 2003, which lists a great amount of thematic fields – without science.

36 Manfred Lachs, "Thoughts on Science, Technology and World Law", *American Journal of International Law* 86 (1992), pp. 673-699.

37 See Armin von Bogdandy/Philipp Dann/Matthias Goldmann (supra note 5), p. 8.

tionalism cannot be left aside by a concept which is aimed at the strengthening of public law mechanisms in the global sphere. The idea of global constitutionalism[38] is indeed focussing upon basic legal principles and structures that are considered to form the foundation of the international community[39]. The core of this concept is the recognition of the international community as an entity with legal relevance, ruled by the aforementioned legal principles and structures, and not just as a conglomerate of States and other subjects of the law[40].

Constitutionalism at global level is not a monolithic theory. A very strong branch of scholars propounding it, do consider the Charter of the United Nations as the constitutive text. Thus, writers like *Alfred Verdross* and *Hermann Mosler* in former times[41] and currently above all *Bardo Fassbender* are detecting constituent elements in the UN Charter[42]: Its constitutional moment after the Second World War, its institu-

---

[38] The discussion whether it is a particularly German concept (Stefan Kadelbach and Thomas Kleinlein, "International Law – a Constitution of Mankind? An Attempt at a Re-appraisal with an Analysis of Constitutional Principles", *German Yearbook of International Law* 50 (2007), pp. 303-347 at p. 304) cannot be dealt with further here.

[39] For a concise overview see Stefan Kadelbach and Thomas Kleinlein (supra note 38).

[40] Andreas Paulus, *Die internationale Gemeinschaft im Völkerrecht,* 2001; Christian Tomuschat, "Die internationale Gemeinschaft", *Archiv des Völkerrechts* 33 (1995), pp. 1-20.

[41] Alfred Verdross, *Die Verfassung der Völkerrechtsgemeinschaft,* 1926 at pp. 12 *et seq.* and 42 *et seq.* in particular (on the League of Nations), and on the same line Alfred Verdross and Bruno Simma, *Universelles Völkerrecht,* 3rd ed. 1984, at VII *et seq.* and para 374; Hermann Mosler, "The International Society as a Legal Community", *Recueil des Cours* 140 (1974-IV), pp. 1 *et seqq.*

[42] The following elements are taken from Bardo Fassbender, "The United Nations Charter As Constitution of the International Community", *Columbia Journal of Transnational Law* 36 (1998), 529. See also Ronald St. John Macdonald, "The United Nations Charter: Constitution or Contract?, in: id. And Douglas M. Johnston (eds.), *The Structure and Process of International Law,* 1983, pp. 889-912; id., "The Charter of the United Nations in Constitutional Perspective", *Australian Yearbook of International Law* 20 (1999), pp. 205-231; Pierre-Marie Dupuy, "The Constitutional Dimension of the Charter of the United Nations Revisited", *Max Planck United Nations Yearbook* 1 (1997), pp. 1-33.; Thomas Franck, "Is the UN Charter a Constitution?", in: Jochen Abraham Frowein (ed.), *Verhandeln für den Frieden, Liber Amicorum Tono Eitel,* 2003, pp. 95-106.

tional framework, its rules on membership, its hierarchical position (Article 103)[43], its stability against revision, its designation ("Charter" instead of treaty or instrument), its role in developing international law and finally its universality. Others are considering the entire international law as the basis for international constitutionalism[44], while still others are stressing the importance of constitutionalising single fields of international law such as international economic law above all[45]. More or less common to all these approaches is the elaboration of the constitutional elements in the field of international law: All authors detect common values, found primarily in the most important human rights but also in central principles and rules of international law (the prohibition of the use of military force, above all)[46], as well as institutional structures which can be developed alongside the classical three-partite differentiation of legislature, executive and judiciary[47].

Global constitutionalism is not a concept beyond contestation. The strongest opposition is formulated because it necessitates the separation between the ideas of constitution and the State. This is still a highly controversial point[48]. Although said separation can be achieved by recognising differences and divergences within the notion of "constitution" itself – the constitution of a State may be a different matter compared to the constitutionalist picture of a multilevel system or the world

---

43 Stefan Kadelbach and Thomas Kleinlein (supra note 38), at pp. 317 *et seq.*

44 For an overview cf. Christian Walter and Matthias Ruffert, *Institutionalisiertes Völkerrecht,* 2009, para. 63.

45 Cf. Ernst-Ulrich Petersmann, *Constitutional functions and constitutional problems of international economic law,* 1991; Peter-Tobias Stoll, "Freihandel und Verfassung", *Zeitschrift für ausländisches öffentliches Recht und Völkerrecht* 57 (1997), pp. 83-146; Markus Krajewski, *Verfassungsperspektiven und Legitimation des Rechts der Welthandelsorganisation,* 2001.

46 Cf. only Bardo Fassbender, "Der Schutz der Menschenrechte als zentraler Inhalt des völkerrechtlichen Gemeinwohls", *Europäische Grundrechte-Zeitschrift* 30 (2003), pp. 1-15.

47 For an overarching approach cf. Christoph Möllers, *Gewaltengliederung,* 2005.

48 Cf. only Rainer Wahl, "Konstitutionalisierung – Leitbegriff oder Allerweltsbegriff?", in: Carl-Eugen Eberle (ed.), *Festschrift Winfried Brohm,* 2002, pp. 191-207 at p. 198. Cf. also the critical assessment by Jan Klabbers, "Constitutionalism Lite", *International Organization Law Review* 1 (2004), pp. 31-58.

at large – it cannot be denied that the difficulty remains that the idea of a constitution which has grown over centuries shall not be flawed[49].

On the basis of its theoretical diversity and bearing in mind the aforementioned caveat, the constitutionalist line in developing international law can be integrated into the concept of global administrative law. The constitutional argumentation provides the link between administrative mechanisms and core public values of public international law as well as a framework for legal institutional analysis. In this respect, global constitutionalism shall be taken up in the analysis of the international governance of science for two reasons: First, it offers the methodological opportunity to cope with the variety of institutional and regulatory levels following the idea of multilevel constitutionalism which integrates the State into a more extensive constitutional field[50]. Multilevel constitutionalism is a common theory to analyse European Union law, as will become apparent in the analysis of EU research policy (below D. III. 2.). Another reason for taking up the constitutionalist view is its focus upon human rights[51]. It is one of the central theses of this book that freedom of science as a human right lies at the very heart of the international governance of science[52].

III. Conclusion

The concept of global administrative law can serve as an analytical tool to strengthen the public law content in public international law[53]. In-

[49] In the context of the European Union: Christoph Möllers, "Pouvoir Constituant – Constitution – Constitutionalisation", in: Armin von Bogdandy and Jürgen Bast (eds.), *Principles of European Constitutional Law,* 2nd ed. 2010, pp. 169-204.

[50] See only Ingolf Pernice, "The Global Dimension of Multilevel Constitutionalism: A Legal Response to the Challenges of Globalisation", in: Pierre-Marie Dupuy/Bardo Fassbender/Malcolm N. Shaw/Karl-Peter Sommermann (eds.), *Völkerrecht als Wertordnung: Festschrift für Christian Tomuschat,* 2006, pp. 973-1005.

[51] Cf. supra note 46. The link is established by Claus-Dieter Classen (supra note 4) at pp. 385 *et seq.*

[52] Cf. infra (C. I.).

[53] Thus, the following study shares the aim of Armin von Bogdandy/Philipp Dann/Matthias Goldmann (supra note 5), at pp. 25 *et seq.,* who do so on a different methodological basis (supra note 5).

stead of a sharp distinction between the ideas of international, global and transnational law, of administrative and constitutional law, the concept represents a comprehensive approach. As in all approaches based on combination, there is, of course, the risk of undue syncretism, of methodologically unsound "cherry picking" without becoming aware of insurmountable divergences between the concepts combined. It can, however, be shown that the risk can be reduced here: Global administrative law can integrate ideas that have so far been dealt with under the title of international administrative law, the concept of transnationality being of both descriptive and systematising value. Global administrative law is also able to take up a constitutional core, and it can be developed around a regulatory reality that is analysed by means of the concept of global governance.

It is with this analytical concept of global administrative law that the field of science shall be opened for international legal scrutiny. It should be noted, in addition, that the particular principles, rules and legal institutions to be detected in the field of science may also have an impact on the design of global administrative law as such. General patterns of a legal area cannot be synthesised, but have to be drawn and developed from particular instances. This interrelationship between the general and the special has been known for a long time in domestic administrative law. Eberhard Schmidt-Aßmann has coined the convincing concept of fields of reference (*Referenzgebiete*), special fields of administrative law with formative effects for administrative law as such[54]. It would certainly be exaggerating to attribute such formative effect to the law of science before even systematising it for the first time. However, general repercussions of what can be found out about the global administrative law of science upon global administrative law as such shall not be overlooked.

---

[54] Eberhard Schmidt-Aßmann, *Das allgemeine Verwaltungsrecht als Ordnungsidee,* 2nd ed. 2004, para. 1/13.

C. Constitutional Basis: The Freedom of Science

I. A Fundamental Right as a Constitutional Basis

Global administrative law enshrines, as developed here, a constitutional basis. The administrative institutions and mechanisms established in the framework of global administrative law to fulfil tasks formulated at global levels are normatively oriented towards constitutional foundations. One of the salient features of the constitutional approach to international law – besides its institutional aspects – is the importance of values embedded in fundamental rules of the international community and in human rights.

On this theoretical basis, we shall develop in this section the idea that freedom of science is a right at the constitutional core of the international governance of science. It is true that the valued rules and rights discussed in the international constitutional field thus far are of more obvious fundamentality, whether we consider the prohibition of the use of force or – to give a clear example – the prohibition of torture and its human rights basis. Nevertheless, the more global administrative law as part of international law extends its reach to diverse fields of application, the more extensive its constitutional background must be. If we consistently follow the constitutional path, it is both possible and necessary to detect and develop rights beneath a level of such absolute fundamentality to weave a more dense constitutional fabric – if only there is a sound method of detection and development.

The method applied here is twofold. Firstly, we will show that freedom of science is present in international treaties and other binding and non-binding instruments. Secondly, we will demonstrate that this very freedom is at the basis of a common constitutional understanding. In considering the relevant treaties and instruments at first, the important rule that no subject of international law is bound by a treaty or instrument it has not adhered to (by ratification, signature or any other emanation of consent)[1] shall not be put aside, but it will become clear that there is an

---

1 Locus classicus: Permanent Court of International Justice, *The Case of the S.S. Lotus*, PCIJ Rep. Ser. A, No. 10, p. 18: "International law governs relations between independent States. The rules of law binding upon States there-

underlying understanding of free research that influences the international governance of science in a constitutional manner. The risk of exaggeration may even be greater in the second part of this section with its comparative orientation. In fact, the size of the risk might be overwhelming if the prominent position of freedom of science in German constitutional law is taken into account[2]. Would the stress upon free research as a constitutional right not come down to a Germanisation of the constitutional footings of the international governance of science? Again, what has to be avoided is the extrapolation of single domestic constitutional guarantees to the international level; instead, the common constitutional idea of the freedom of science has to be elaborated together with its constitutional effect on the international governance of science.

II. Freedom of Science in International Law

1. Universal Human Rights Instruments

The textual presence of freedom of academic research in universal human rights instruments is rather scarce, but it would also be incorrect to deny its importance. Article 27 (1) of the Universal Declaration of Human Rights (1948) reads:

> "Everyone has the right freely to participate in the cultural life of the community, to enjoy the arts and to *share in scientific advancement* and its benefits."[3]

It is of course difficult to infer from this article alone that there is a universal human right to perform free scientific research[4]. This is less due

---

fore emanate from their own free will as expressed in conventions or by usages generally accepted as expressing principles of law and established in order to regulate the relations between these coexisting independent communities or with a view to the achievement of common aim. Restrictions upon the independence of States cannot therefore be presumed."

2 Cf. Thomas Oppermann, "Freiheit von Forschung und Lehre", in: Josef Isensee and Paul Kirchhof (eds.), *Handbuch des Staatsrechts, Band VI Freiheitsrechte*, 2001, pp. 809-845, paras. 62 *et seq.*

3 993 U.N.T.S. 3 (emphasis by the authors).

4 On the history of the provision see Richard Claude, "Scientists' Rights and the Human Right to the Benefits of Science", in: Audrey Chapman (ed.),

to the Declaration's legal status, which is considered by a majority of writers to have crystallised into binding international law[5]. What is more important is the actual formulation of the article. It does not even mention the scientist, but concentrates upon the beneficiaries of science in a certain way. What is at stake is the free access to the results of academic research[6]. It has to be noted, though, that the most fundamental document of post-1945 human rights development mentions "scientific advancement" as a core resource access to which must be guaranteed.

Article 15 (3) of the International Covenant on Economic, Social and Cultural Rights (1966) may be more relevant:

> "The States Parties to the present Covenant undertake to respect the freedom indispensable for scientific research and creative activity."

To begin with, the Covenant is an international treaty the binding force of which is beyond doubt – the complicated mechanism of its implementation notwithstanding[7]. But the wording of the text can be considered as the conferment of an individual right only under great difficulties [8], particularly if its Article 15 (1) is taken into account, which reads:

> "The States Parties to the present Covenant recognize the right of everyone: (a) To take part in cultural life; (b) *To enjoy the benefits of scientific progress and its applications*; (c) To benefit from *the protec-*

Core obligations: Building a framework for political, social and cultural rights, 2002, pp. 247-278 at pp. 251 *et seq.*

[5] See the clarification by Antonio Cassese, *International Law,* 2nd ed. 2005, pp. 380-382. Cf. Also Ian Brownlie, *Principles of Public International Law*, 7th ed. 2008, pp. 559 *et seq.*

[6] Cf. Thomas Groß, *Die Autonomie der Wissenschaft im europäischen Rechtsvergleich,* 1992, p. 177. In this context, see also the Declaration on the Use of Scientific and Technological Progress in the Interests of Peace and for the Benefit of Mankind, GA Res. 3384 (XXX) of 10 November 1975.

[7] Ian Brownlie (supra note 5), pp. 565 *et seq*; Kathleen Renée Cronin-Furman, "60 Years of the Universal Declaration of Human Rights: Towards an Individual Responsibility to Protect", *The American University International Law Review* Vol. 25 Issue 1 (2009) pp. 175-198 at pp. 184 *et seqq.*

[8] Thomas Groß (supra note 6), p. 178. Cf. also the duty of the State Parties to progressively realise the rights recognised in the Covenant in Article 2 (1); Michael Fehling, in: Rudolf Dolzer/Karin Graßhof/Wolfgang Kahl/Christian Waldhoff (eds.), *Bonner Kommentar zum Grundgesetz*, 148th ed. 2010, Article 5 (3), para. 278.

> *tion of the moral and material interests* resulting from any *scientific*, literary or artistic *production* of which he is the author."[9]

Again, as in the Universal Declaration, the benefits of scientific progress are underlined and transferred into a social context[10]. New here, however, is the right to protection of intellectual property for scientists also [11]. It may be due to the textual scarcity that the United Nations Committee on Economic, Social and Cultural Rights (UNCESCR), the Committee erected to implement the Covenant, treated freedom of scientific research as a special category of the right to education:

> "Members of the academic community, individually or collectively, are free to pursue, develop and transmit knowledge and ideas, through research, teaching, study, discussion, documentation, production, creation or writing."[12]

This, however, appears imprecise as the concept of science transcends the educative sphere[13].

Freedom of science does not appear as a basic human right in the International Covenant on Civil and Political Rights (1966). However, an aspect of that freedom is regularly mentioned in the reports of the Special Rapporteurs on the freedom of expression: the free publication of research results and academic writings[14]. Consequently, it cannot be

---

[9] Emphasis by the authors.

[10] Claude (supra note 4), at p. 255, underlines the egalitarian aspect of this provision.

[11] On this Maria Green, *Drafting History of the Article 15 (1) (c) of the International Covenant on Economic, Social and Cultural Rights,* Background Paper, UN Doc. E/C.12/2000/15 (9 October 2000), at Nr. 45.

[12] Committee on Economic, Social and Cultural Rights (CESCR), General Comment 13: The Right to Education, E/C.12/1999/10, 1999, Nr. 39; Committee on Economic, Social and Cultural Rights, *Substantive Issues Arising in the Implementation of the International Covenant on Economic, Social and Cultural Rights,* UN Doc. E/C.12/2001/15 (14 December 2001).

[13] Sebastian Steinecke, *Zur internationalen Governance der Wissenschaft,* 2010, manuscript pp. 277 *et seq.*

[14] Vgl. Abid Hussain, *Civil and Political Rights Including the Question of Freedom of Expression,* E/CN.4/2000/63 of 18th January 2000. Although this admission is not consequently done; in the subsequent reports until 2010 the Freedom of Science is not longer mentioned separately. According to the cited example graduates are apparently only noticed in their role as active citizens not as scientists.

maintained that the absence of any textual reference to that right would signify that the signatory States of the Covenant were opposed to its recognition.

Nevertheless, the feeble entrenchment of freedom of science cannot be denied[15]. Instead of a clear, overall guarantee, we can detect elements in different texts and institutional statements. Three elements should be borne in mind: (1) the recognition of the beneficial effect of scientific progress, (2) the freedom to publish results and research opinions and (3) the protection of the scientists' intellectual property.

2. Regional Human Rights Treaties

With few exceptions, regional human rights treaties do not provide a guarantee of the human right considerably more strengthened. Nevertheless, the abovementioned elements of a guarantee at least partially appear in these treaties. Although neither the European Convention on Human Rights (1950)[16], nor the American Convention on Human Rights (1978)[17], nor the African Charter of Human and Peoples' Rights (1986)[18] do contain the right to free scientific research explicitly, it is recognised that at least the free publication of scientific texts is part of the freedom of expression[19].

---

15 Cf. also Matthias Ruffert, "Grund und Grenzen der Wissenschaftsfreiheit", *Veröffentlichungen der Vereinigung der Deutschen Staatsrechtslehrer* 65 (2006), pp. 145-210 at p. 169.

16 213 U.N.T.S. 221.

17 1144 U.N.T.S. 143. Nonetheless, the American Declaraton of the Righs and Duties of Man (1948), provides in its Article 13: "(1) Every person has the right to ... participate in the benefits that result from intellectual progress, especially scientific discoveries..." and "(2) scientists have a right to protection of their moral and material interests for their scientific works" (American Declaration of the Rights and Duties of Man, Resolution XXX, *Novena Conferencia Internacional Americana, Actas y Documentos, vol. VI*, Bogotá 1953, pp. 297-302).

18 1520 U.N.T.S. 217.

19 Cf. Sebastian Steinecke (supra note 13), manuscript pp. 285 *et seq.*

Astonishingly enough, it is the Arab Charter of Human Rights (2008)[20] that contains a relevant provision in its Article 42 Nr. 2:

> "The States parties undertake to respect the freedom of scientific research and creative activity and to ensure the protection of moral and material interests resulting from scientific and artistic production."[21]

The effective reach of that Charter cannot easily be assessed, though:

Therefore, the second exception to textual abstention is far more important. Following Article 6 (1) of the Treaty on European Union (TEU) as in force after the modifications of the Treaty of Lisbon[22], the Charter of Fundamental Rights in the European Union[23] as proclaimed in Nice in 2000 and re-proclaimed in Strasbourg in 2007 is a binding human rights instrument at EU level[24]. Its Article 13 provides:

> "The arts and scientific research shall be free of constraint. Academic freedom shall be respected."

This common European standard[25] is indispensable for the establishment of free research given the broad EU powers in this field (below D. III. 2.). It should be noted that the Charter is basically binding upon the EU; the Member States are only bound in a restricted manner (Article 51 (1) of the Charter)[26]. A certain difficulty in interpreting the newly

[20] Reprinted in International Human Rights Report 12 (2005), pp. 893 *et seqq.*; text and translation available at University of Minnesota – Human Rights Library http://www1.umn.edu/humanrts/instree/loas2005.html?msource=UNWDEC19001&tr=y&auid=3337655.

[21] Cf. supra note 20.

[22] Treaty of Lisbon amending the Treaty on European Union and the Treaty establishing the European Community, signed at Lisbon, 13 December 2007, [2007] O.J. C 306/1.

[23] [2010] O.J. C 83/389.

[24] See Damian Chalmers/Gareth Davies/Giorgio Monti, *European Union Law*, 2nd ed. 2010, pp. 230 *et seq.*

[25] This is expressly underlined by Ginamario Demuro, "Article 13", in: William B.T. Mock/id./Raffaele Bifulco/Marta Cartabia/Alfonso Celotto (eds.), *Human Rights in Europe*, 2010, pp. 84-87 at pp. 85 *et seqq.* Cf. also Matthias Ruffert, in: Christian Calliess and id. (eds.), *EUV/AEUV*, 4th ed. 2011, Article 13 GRCh, para. 1.

[26] See Christian Calliess, "The Charter of Fundamental Rights of the European Union", in: Dirk Ehlers (ed.), *European Fundamental Rights and Freedoms*, 2007, paras. 20/25 *et seq.*

established right is the risk of interpreting it only in the context of the long existing German fundamental right. Further, academic freedom (second sentence of Article 13) is concomitant to free scientific research in the universitary world[27]. All in all, Article 13 of the Charter of Fundamental Rights is not only important to delimit EU powers for the benefit of free research but also as a source for the development of a human rights driven international governance of science.

3. Further Binding Instruments

In contrast to the relative scarcity of science provisions in universal and regional human rights treaties, the subject is quite often treated in other binding multi- or bilateral international instruments. Such instruments may contain the issue of research in particular clauses. Of course, they do not entail the explicit recognition of a human right to free scientific research – that would have been the subject matter of a treaty of its own[28]. The content of such agreements, however, shows that the State parties adhere to a concept of science which would not be viable if free research was not legally guaranteed.

Elmar Döhler and *Carsten Nemitz* have shown that such treaty-clauses on research can be categorised under five headlines: purposes (such as international co-operation, exchange of information, information of the general public and dissemination of research outcome), support of science including privileges for science, limitations for the preservation of opposing values institution building and two other issues[29]. The diverse means of supporting science are the most important category identified by these authors[30]. What follows is that if science is to be supported and opposing values (and rights) are to be protected against some of its outcomes, there is an implicit recognition of the necessity of free scientific

---

27 Matthias Ruffert (supra note 25), Article 13 GRCh, para. 9.

28 See the explicit criticism by Elmar Döhler and Carsten Nemitz, "Wissenschaft und Wissenschaftsfreiheit in internationalen Vereinbarungen", in: Hellmut Wagner (ed.), *Rechtliche Rahmenbedingungen für Wissenschaft und Forschung. Forschungsfreiheit und Staatliche Regulierung*, Vol. 1, 2000, pp. 159-188, at pp. 182-185.

29 Elmar Döhler and Carsten Nemitz (supra note 28).

30 Elmar Döhler and Carsten Nemitz (supra note 28), at pp. 173 *et seq.*

activity. The benefits of science shall be freely accessible (function of information).

In the context of support and co-ordination, several multilateral treaty provisions have to be mentioned. First of all, Articles 238 *et seq.* (i.e. Part XIII) of the United Nations Convention on the Law of the Sea (1994)[31] contain extensive provisions about marine scientific research. The core general provisions are as follows:

> Article 238 Right to conduct marine scientific research
>
> All States, irrespective of their geographical location, and competent international organizations have the right to conduct marine scientific research subject to the rights and duties of other States as provided for in this Convention.
>
> Article 239 Promotion of marine scientific research
>
> States and competent international organizations shall promote and facilitate the development and conduct of marine scientific research in accordance with this Convention.
>
> Article 240 General principles for the conduct of marine scientific research
>
> In the conduct of marine scientific research the following principles shall apply:
>
> (a) marine scientific research shall be conducted exclusively for peaceful purposes;
>
> (b) marine scientific research shall be conducted with appropriate scientific methods and means compatible with this Convention;
>
> (c) marine scientific research shall not unjustifiably interfere with other legitimate uses of the sea compatible with this Convention and shall be duly respected in the course of such uses;
>
> (d) marine scientific research shall be conducted in compliance with all relevant regulations adopted in conformity with this Convention

---

[31] 1833 U.N.T.S. 3. Cf. Volker Röben, "The Sciences – A Contribution to Understanding the Law on an Activity of International Concern", *German Yearbook of International Law* 37 (1994), pp. 254-280 at p. 258; Wolf Plesmann and Volker Röben, "Marine Scientific Research: State Practice versus Law of the Sea?", in: Rüdiger Wolfrum (ed.), *Law of the sea at the crossroads: the continuing search for a universally accepted régime*, 1991, pp. 373-392, in detail Myron H. Nordquist/Alexander Yankov/Neal R. Grandy/Shabtai Rosenne, *United Nations Convention on the Law of the Sea 1982 – A Commentary, Vol. IV*, 2002, pp. 429 *et seqq*.

including those for the protection and preservation of the marine environment.

...

Article 242 Promotion of international cooperation

1. States and competent international organizations shall, in accordance with the principle of respect for sovereignty and jurisdiction and on the basis of mutual benefit, promote international cooperation in marine scientific research for peaceful purposes. ...

Article 243 Creation of favourable conditions

States and competent international organizations shall cooperate, through the conclusion of bilateral and multilateral agreements, to create favourable conditions for the conduct of marine scientific research in the marine environment and to integrate the efforts of scientists in studying the essence of phenomena and processes occurring in the marine environment and the interrelations between them.

Article 244 Publication and dissemination of information and knowledge

1. States and competent international organizations shall, in accordance with this Convention, make available by publication and dissemination through appropriate channels information on proposed major programmes and their objectives as well as knowledge resulting from marine scientific research.

2. For this purpose, States, both individually and in cooperation with other States and with competent international organizations, shall actively promote the flow of scientific data and information and the transfer of knowledge resulting from marine scientific research, especially to developing States, as well as the strengthening of the autonomous marine scientific research capabilities of developing States through, inter alia, programmes to provide adequate education and training of their technical and scientific personnel.

What is pertinent in these provisions is the obvious intention to promote research either by States alone, by International Organisations or by other co-operative structures. Articles 245-265 of the Convention contain detailed rules on the performance of scientific research projects in line with the overall concept of UNCLOS, including responsibility/liability and dispute settlement.

Articles II, III, IX (1) (b) and (c) of the Antarctic Treaty (1959)[32] expressly underline the freedom of scientific research in Antarctica and provide for co-operative performance of that research:

Article II

Freedom of scientific investigation in Antarctica and cooperation toward that end, as applied during the International Geophysical Year, shall continue, subject to the provisions of the present treaty.

Article III

1. In order to promote international cooperation in scientific investigation in Antarctica, as provided for in Article II of the present treaty, the Contracting Parties agree that, to the greatest extent feasible and practicable:

(a) information regarding plans for scientific programs in Antarctica shall be exchanged to permit maximum economy and efficiency of operations;

(b) scientific personnel shall be exchanged in Antarctica between expeditions and stations;

(c) scientific observations and results from Antarctica shall be exchanged and made freely available.

2. In implementing this Article, every encouragement shall be given to the establishment of cooperative working relations with those Specialized Agencies of the United Nations and other international organizations having a scientific or technical interest in Antarctica.

Article IX

1. Representatives of the Contracting Parties named in the preamble to the present treaty shall meet at the City of Canberra within two months after the date of entry into force of the treaty, and thereafter at suitable intervals and places, for the purpose of exchanging information, consulting together on matters of common interest pertaining to Antarctica, and formulating and considering, and recommending to their Governments, measures in furtherance of the principles and objectives of the treaty, including measures regarding:

...

(b) facilitation of scientific research in Antarctica;

---

[32] 402 U.N.T.S. 71. Cf. Volker Röben (supra note 31) at p. 258; Helmut Wagner, "Gibt es ein Grundrecht der Wissenschaftsfreiheit im Europäischen Gemeinschaftsrecht", *Die öffentliche Verwaltung* 1999, pp. 129-137 at p. 134.

(c) facilitation of international scientific cooperation in Antarctica;

Article 1 (3) of the Treaty on the Principles Governing the Activities of States in the Exploration and Use of Outer Space, Including the Moon and Other Celestial Bodies (1967)[33] is drafted on similar lines,

> There shall be freedom of scientific investigation in outer space, including the moon and other celestial bodies, and States shall facilitate and encourage international co-operation in such investigation.

Finally, a number of international environmental agreements and a very particular agreement provide for the promotion of scientific research by means of co-operative structures and own material efforts: Article 4 of the Vienna Convention for the Protection of the Ozone Layer (1988)[34], Article 12 of the Convention on Biological Diversity (1993)[35], besides, even Article 9 (3) United Nations Convention Against Illicit Traffic in Narcotic Drugs and Psychotropic Substances (1990)[36].

4. Instruments of International Organisations

Declarations and other instruments issued by International Organisations take up elements that have already been mentioned in the context of the human rights treaties, but there are instances in which the tendency to achieve the guarantee of free scientific research is stronger than in binding instruments.

Along this line, the General Assembly continued to emphasise the beneficiary aspect of scientific research as it did in the 1948 human rights declaration. In its Declaration on the Use of Scientific and Technological Progress in the Interests of Peace and for the Benefit of Mankind[37], the General Assembly proclaimed:

> "All states shall promote international co-operation to ensure that the results of scientific and technological developments are used in the interests of strengthening international peace and security, freedom and independence, and also for the purpose of the economic and social development of peoples ..."

---

33 610 U.N.T.S. 205.

34 323 U.N.T.S. 1513.

35 1760 U.N.T.S. 79.

36 UN Doc. E/CONF.82/15 (I.L.M. 28 (1989), pp. 493 *et seqq.*).

37 UN Doc. A/RES/30/3384 (10 November 1975).

It took until 1998 when the General Assembly endorsed the UNESCO Universal Resolution on the Human Genome and Human Rights[38] which contains, in its Article 12 (b) the following passage:

> "Freedom of research, which is necessary for the progress of knowledge, is part of freedom of thought. The applications of research, including applications in biology, genetics and medicine, concerning the human genome, shall seek to offer relief from suffering and improve the health of individuals and humankind as a whole."

Along the same lines, it is one of the aims endorsed in Article 2 of the Universal Declaration on Bioethics and Human Rights (2005)

> "... (d) *to recognize the importance of freedom of scientific research and the benefits derived from scientific and technological developments*, while stressing the need for such research and developments to occur within the framework of ethical principles set out in this Declaration and to respect human dignity, human rights and fundamental freedoms;..."[39]

It is submitted that statements bearing that content, issued in an extremely controversial field (cf. below E. II. 2.) can provide an indication for the current meaning of freedom of science. Along the same line, the World Bank, in a Report of 2000, defined academic freedom as

> "... the right of scholars to pursue their research, to teach, and to publish without control or restraint from the institutions that employ them. (...) Academic freedom is not an absolute concept; it has limits and requires accountability. It recognizes the right of academics to define their own areas of inquiry and to pursue the truth as they see it."[40]

Again, freedom of science is emphasised as an individual right, but limits and accountability are equally stressed.

---

[38] UNESCO, *Records of the Twenty-Ninth Session*, 1997, 29 C/Res. 16 (11 November 1997), endorsed by UN Doc. A/RES/53/152 (9 December 1998).

[39] UNESCO Doc. C/Res. 24 of 19 October 2005, emphasis by the authors. On that document see Harald Schmidt, "Bioethics, Human Rights and Universalisation: a Troubled Relationship? – Observations on UNESCO's Universal Declaration on Bioethics and Human Rights", in: Silja Vöneky/Cornelia Hagedorn/Miriam Clados/Jelena von Achenbach (eds.), *Legitimation ethischer Entscheidungen im Recht*, 2009, pp. 275-295.

[40] The World Bank (ed.), *Higher Education in Developing Countries*, 2000, p. 60. The definition was taken without amendment from Barbara Chernow (ed.), "Art. Academic Freedom", *The Columbia Encyclopedia*, 5th ed. 1993.

5. Instruments of Non-Governmental Organisations

The great variety of non-governmental organisations (NGOs) in the field of science has yet to be shown (below D. IV. 2.). They are also partly active in formulating general standards for the international governance of science. Thus, the Lima Declaration on Academic Freedom and Autonomy of Institutions of Higher Education (1988) is revelatory. It was issued by the World University Service and provides for a broad guarantee of freedom of science:

> "'Academic freedom' means the freedom of members of the academic community, individually or collectively, in the pursuit, development and transmission of knowledge, through research, study, discussion, documentation, production, creation, teaching, lecturing and writing."

As an NGO to support university studies, the WUS is of course not entitled to issue legally binding statements in any respect, but the content of the declaration is illustrative. Freedom of science cannot be limited to one area or another of scientific activity, but it needs a more comprehensive protection.

6. Contents and Effective Potential of the International Standard

The overall impression gained from an analysis of existing international rules including rules of doubtful binding force or of certainly non-binding nature is ambiguous. On the one hand, there is rarely a clear guarantee of freedom of science in an at least regionally binding document; Article 13 of the EU's newly binding human rights charter is certainly the strongest emanation of such a right. On the other hand, some elements are included into the various instruments which delineate a picture of the idea of a global freedom of science and which also convey the impression of continuous progress in the establishment of such a freedom, starting from the mere recognition of the benefits of (free) science, encompassing free publication and the ownership of intellectual property and finally ending up in definitions including a broad range of scientific activities.

III. Freedom of Science in National Constitutions

1. Methodological Remarks

In search of global constitutional rights and values, the scrutiny of existing public international legal principles and rules is not the only viable method, although of course there is little reasoning other than the detection of whatever value within positive public international law that would be of overall convincing evidence. The inclusion of domestic constitutional law therefore needs a sound methodological reflection. As far as the content of the constitutional guarantees is concerned, it will be shown that all divergences in contents notwithstanding, there is widespread recognition of at least some elements of the guarantee of freedom of science.

The central public international legal norm to integrate aspects of domestic constitutional law into public international law is Article 38 (1) (c) of the Statue of the International Court of Justice. Certainly, "... the general principles of law recognised by civilized nations; ..." were meant to be core principles of the rule of law, and thus very often private law rules, in the times of public international law as a law of co-ordination. The constitutional shift of public international law however is dependent upon the modification of that meaning. If public international law is built upon common constitutional values, their recognition has to be achieved even if such values are – for whatever reason – basically entrenched in domestic constitutions and not in international instruments.

The doubtful, historically rooted term "civilized nations" put aside, it is obvious that the formulation of a general principle, in particular if constitutional value is ascribed to it, necessitates overwhelming if not unanimous recognition within the constitutional texts under scrutiny[41]. This necessity raises the central problems of efforts in comparative law that go beyond the mere comparison of a limited number of jurisdictions: The more jurisdictions are integrated into an analysis, the smaller the relation between effort and outcome, and the greater the difficulties

[41] Frances T. Freeman Jalet, "The Quest for the General Principles of Law Recognized by Civilized Nations – a Study", *UCLA Law Review* 10 (1962-1963), pp. 1041-1086 at pp. 1044 *et seqq.*; Béla Vitanyi, "Les positions doctrinales concernant le sens de la notion de 'principes généraux de droit reconnus par les nations civilisées'", *Revue Générale de Droit International Public* 86 (1), 1982, Paris, pp. 48-116.

in gathering valid material due to linguistic deficiencies that are not easily overcome[42].

2. Constitutional Provisions

a) Categorisation

These methodological reflections and limitations borne in mind, the constitutional picture is manifold and not free of controversies, above all if the internal practice and academic discussions within the various jurisdictions are taken into account. For an easier assessment it is necessary to categorise the different constitutional provisions, albeit with the disadvantage of simplifications. Three main groups can be considered:

b) Overall Guarantee

The constitutions of a considerable number of States of the world plainly guarantee free scientific activity and academic freedom. This applies for most States of the EU, laying a basis for the fundamental right as guaranteed in the EU's Human Rights' Charter: Austria[43], Bulgaria[44], Czech Republic[45], Estonia[46], Finland[47], Hungary[48], Italy[49], Germany,

---

[42] Cf. Matthias Ruffert, "The Transformation of Administrative Law as a Transnational Methodological Project", in: id. (ed.), *The Transformation of Administrative Law in Europe – La mutation du droit administrative en Europe,* 2007, pp. 3-52 at p. 9.

[43] Article 17 (1) of the Basic Law on the General Rights of Nationals of the Republic of Austria (1867): "*Knowledge and its teaching are free.*" (translated by the Federal Chancellery, available at http://www.ris.bka.gv.at/Dokument.wxe?Abfrage=Erv&Dokumentnummer=ERV_1867_142&ResultFunctionToken=1c7a0cca-bd32-41f6-a6c1-984e4e34c8ff&Titel=Basic+Law+on+the+General+Rights+of+Nationals+%281867%29&Quelle=&ImRisSeit=Undefined&ResultPageSize=50&Suchworte).

[44] Article 54 (2) Constitution of the Republic of Bulgaria (1991): "*Artistic, scientific and technological creativity shall be recognized and guaranteed by the law.*" (source: the National Assembly of the Republic of Bulgaria; available at http://www.parliament.bg/?page=const&lng=en).

[45] Article 15 (1) and (2) of the Charter of Fundamental Rights and Basic Freedoms (1992): "*(1) The freedom of thought, conscience, and religious conviction is guaranteed. Everyone has the right to change her religion or faith or to be non-denominational. (2) The freedom of scholarly research and of artistic crea-*

Greece[50], Latvia[51], Lithuania[52], Poland[53], Portugal[54], Slovakia[55], Slove-

---

tion is guaranteed.". Art. 3 of the Constitution of the Czech Republic (1992): "*An integral component of the constitutional system of the Czech Republic is the Charter of Fundamental Rights and Freedoms.*" (source: the Parliament of the Czech Republic; available at http://www.psp.cz/cgi-bin/eng/docs/laws/1993/2.html).

46 Article 38 Constitution of the Republic of Estonia (1992): "*[Freedom of Science and Teaching] (1) Science and the arts, and their instruction, shall be able to exist freely. (2) Universities and research institutions shall be autonomous, within the limits prescribed by law.*" (source: the President of the Republic of Estonia; available at http://www.president.ee/en/estonia/constitution.php).

47 Section 16 Constitution of the Republic of Finland (1999): "*Educational rights: (1) Everyone has the right to basic education free of charge. Provisions on the duty to receive education are laid down by an Act. (2) The public authorities shall, as provided in more detail by an Act, guarantee for everyone equal opportunity to receive other educational services in accordance with their ability and special needs, as well as the opportunity to develop themselves without being prevented by economic hardship. (3) The freedom of science, the arts and higher education is guaranteed.*" (translated by the Ministry of Justice; available at http://www.finlex.fi/en/laki/kaannokset/1999/en19990731?search[type]=pika&search[pika]=Constitution).

48 Article 70G Constitution of the Republic of Hungary (1949): "*(1) The Republic of Hungary shall respect and support the freedom of scientific and artistic expression, the freedom to learn and to teach.(2)Only scientists are entitled to decide in questions of scientific truth and to determine the scientific value of research.*" (source: the Constitutional Court of the Republic of Hungary; available at http://www.mkab.hu/index.php?id=constitution).

49 Article 33 Constitution of the Italian Republic (1947): "*[Freedom of Arts, Science and Teaching] (1) The arts and sciences as well as their teaching are free.*" (source: the Chamber of Deputies; available at http://legxven.camera.it/cost_reg_funz/345/346/listaArticoli.asp).

50 Article 16 Constitution of Greece (1975) "*[Education]: (1) Art and science, research, and teaching are free and their development and promotion constitutes a state obligation. Academic freedom and the freedom to teach do not override the duty to obey the Constitution.*" (translated by *George Katrougalos*; available at http://www.servat.unibe.ch/icl/gr00000_.html).

51 Article 113 Constitution of Latvia (1922): "*[Research, Art, Copyright, Patents] The State shall recognise the freedom of scientific research, artistic and other creative activity, and shall protect copyright and patent rights.*" (translation by the Latvia Law Institute, available at http://www.humanrights.lv/doc/latlik/satver~1.htm).

nia[56] and Spain[57]. Other EU Member States provide a guarantee at statutory level[58]. The German guarantee is very well developed in the juris-

---

52 Article 42 Constitution of the Republic of Lithuania (1992): "*Culture, science and research, and teaching shall be free. The State shall support culture and science, and shall take care of the protection of Lithuanian historical, artistic and cultural monuments and other culturally valuable objects. The law shall protect and defend the spiritual and material interests of an author which are related to scientific, technical, cultural, and artistic work.*" (source: the Constitutional Court of the Republic of Lithuania; available at http://www.lrkt.lt/Documents2_e.html).

53 Article 73 Constitution of the Republic of Poland (1997): "*The freedom of artistic creation and scientific research as well as dissemination of the fruits thereof, the freedom to teach and to enjoy the products of culture, shall be ensured to everyone.*" (translated by Albert Pol and Andrew Caldwell; available at http://www.sejm.gov.pl/prawo/konst/angielski/kon1.htm).

54 Article 42 Constitution of Portugal (1976): "*Freedom of Cultural Creation: (1) Intellectual, artistic, and scientific creation are unrestricted. (2) This freedom includes the right to invention, production, and dissemination of scientific, literary, or artistic works, including legal protection of copyright.*" (source: the Assembly of the Republic of Portugal; available at http://app.parlamento.pt/site_antigo/ingles/cons_leg/Constitution_VII_revisao_definitive.pdf).

55 Article 43 Constitution of the Slovak Republic (1993): "*(1) Freedom of scientific research and in art are guaranteed. The rights to the results of creative intellectual activity are protected by law. (2) The right of access to the cultural heritage is guaranteed under conditions defined by law.*" (source: Government Office of the Slovak Republic; available at http://www.vlada.gov.sk/9717/part-2-fundamental-rights-and-freedoms.php).

56 Article 59 Constitution of the Republic of Slovenia (1991): "*(Freedom of Science and the Arts) The freedom of scientific and artistic endeavour shall be guaranteed.*" (source: the National Council of the Slovak Republik; available at http://www.us-rs.si/o-sodiscu/pravna-podlaga/ustava/?lang=1).

57 Article 20 Constitution of Spain (1978): "*[Specific Freedoms, Restrictions] (1) The following rights are recognized and protected: (b) Literary, artistic, scientific, and technical production, and creation. (c) Academic freedom.*" (source: the Spanish Senate (Upper House); available at http://www.senado.es/constitu_i/index.html).

58 Cf. France: Law on Higher Education (1983), available (in French) at http://admi.net/jo/loi84-52.html; The Netherlands: The Higher Education and Scientific Research Act (1992), available (in Dutch) at http://www.win.tue.nl/reglementen/whw26092003.pdf; The United Kingdom: Education Reform Act (1988), available at http://www.legislation.gov.uk/ukpga/1988/40/contents. Cf. further Thomas Mann, "Forschungsfreiheit und akademische Freiheit", in: F. Sebastian M. Heselhaus and Carsten Nowak (eds.), *Handbuch der Europäi-*

prudence of the Federal Constitutional Court (*Bundesverfassungsgericht*) with a particular emphasis on the institutional guarantee of universities following the *Humboldtian* ideal[59], although this guarantee has been weakened in more recent judgments[60].

None of these guarantees is unlimited, of course: Like all human rights, freedom of science is subject to limitations necessary in democratic societies. Those limitations are valid, if only they are founded in valid legislation and – in most jurisdictions – are proportionate.

Outside the EU, there are several constitutional jurisdictions providing similar guarantees. Though in Japan the proximity to freedom of speech (cf. below c.) is emphasised[61] by some authors, Article 23 of the Japanese Constitution (1946) is similar in its actual wording and intellectually linked to the comprehensive German guarantee (cf. above fn. 59)[62]. The Russian Constitution (1993) provides in its Article 44 (1):

> "Everyone shall be guaranteed the freedom of literary, artistic, scientific, technical and other types of creative activity, and teaching"[63].

---

schen Grundrechte, 2006, pp. 745-770 paras. 37 *et seqq.*; Thomas Groß (supra note 6).

59 *Entscheidungen des Bundesverfassungsgerichts* 15, 256; 35, 79 (leading case); 47, 327; 51, 369; 55, 37; 85, 360; 88, 129; 93, 85; 111, 333.

60 Particularly in *Entscheidungen des Bundesverfassungsgerichts* 111, 333.

61 Ken Nemori, "Grenzen der Wissenschafts- und Forschungsfreiheit in der japanischen Verfassung und das Klonen von Menschen", in: Rainer Wahl (ed.), *Verfassungsrecht der Humangenetik im deutsch-japanischen Vergleich*, 2002, pp. 224-255 at p. 233, available at: http://www.freidok.uni-freiburg.de/volltexte/5029/pdf/Verfassungsrecht_der_Humangenetik_im_deutsch_japanischen_Vergleich.pdf.

62 Article 23 Constitution of Japan (1946): "*Academic freedom is guaranteed.*" On this provision and its relationship to the German guarantee see Kenichi Moriya, "Wissenschaftsfreiheit. Beobachtungen zum deutschen und japanischen juristischen Diskurs", *Rechtsgeschichte* 7 (2005), pp. 74-85.

63 English translation (by "Garant-Service") of the Russian Constitution available at http://www.constitution.ru/en/10003000-01.htm. Cf. on this Article the Decision of the Plenum of the Supreme Court of the Russian Federation No. 15 of 19 June 2006 on the Issues Emerging with the Courts When Processing Civil Cases Pertaining to Application of the Legislation on the Copyright and Adjacent Rights, available (in Russian) at http://www.supcourt.ru/vscourt_detale.php?id=4349.

Nominally, the guarantee is as wide as in other European countries[64]. Impairment of this freedom is suffered in particular due to the insecure economic situation[65].

Finally, there are States which nominally provide for a comprehensive guarantee of freedom of science, but where there is only limited benefit from such guarantee in constitutional practice. It is with this limitation that Article 47 of the Constitution of the People's Republic of China (1982) must be read:

> "Citizens of the People's Republic of China have the freedom to engage in scientific research, literary and artistic creation and other cultural pursuits. The state encourages and assists creative endeavors conducive to the interests of the people that are made by citizens engaged in education, science, technology, literature, art and other cultural work."[66]

In Saudi Arabia, Islamic legal rules and principles are superimposed upon the constitutional guarantee in Article 29 of the Constitution (1992):

> "The state safeguards science, literature and culture; it encourages scientific research; it protects the Islamic and Arab heritage and contributes toward the Arab, Islamic and human civilization."[67]

Nonetheless, in both of these States, science is of major importance – above all to economic development – and Saudi Arabia even undertakes considerable efforts to attract foreign scientists[68].

---

64 Anna Smolentseva, "Challenges to the Russian academic profession", *Higher Education* 45 (2003), pp. 391-424 at p. 417.

65 Brian Levin-Stankevich and Alexander Savelyev, "The Academic Profession in Russia", in: Philip Altbach, (ed.), *The International Academic Profession. Portraits of Fourteen Countries,* 1996, pp. 569-614 at pp. 590 *et seq.*

66 Cf. also its Article 20: "*The state promotes the development of the natural and social sciences, disseminates scientific and technical knowledge, and commends and rewards achievements in scientific research as well as technological discoveries and inventions.*" English translation available at http://www.gov.cn/english/ 2005-08/05/content_20813.htm (Source: China Yearbook 2004).

67 Cf. English translation: Axel Tschentscher (ed.), *International Constitutional Law*, available at http://www.servat.unibe.ch/icl/sa00000_.html.

68 Sebastian Steinecke (supra note 13), manuscript p. 241.

c) Academic Freedom and the Freedom of Speech

Other constitutions do not contain a textual reference to the right of free scientific research, but often elements of that right as mentioned in the context of international instruments above are guaranteed. This is particularly true for the United States of America. In the U.S., freedom of research and academic freedom are considered to be an aspect of the freedom of speech as established in the First Amendment (1791). The leading case is *Sweezy* v. *State of New Hampshire,* which concerned the Marxist contents of the courses of a guest lecturer and their potential prohibition in the context of the strict legislation (and atmosphere) of the 1950s and where it was held that:

> "Petitioner's right to lecture and his right to associate with others were constitutionally protected freedoms ... Teachers and students must always be free to inquire, to study and to evaluate, to gain new maturity and understanding. ..."[69].

This line of jurisprudence was later continued in other decisions such as the one in *Griswold* v. *Connecticut* where it was held that:

> "The right of freedom of speech and press includes ... freedom of inquiring, freedom of thought and freedom to teach."[70]

Later cases extended the guarantee to the whole universitary field[71] and underlined the importance of academic freedom within the freedom of speech:

> "Our nation is deeply committed to safeguarding academic freedom, which is a transcendent value to all of us and not merely to the teachers concerned. That freedom is therefore a special concern of the First Amendment, ..."[72]

or

---

69 354 U.S. 234 (1957), at p. 250.

70 381 U.S. 479 (1965), at p. 482.

71 *Barenblatt* v. *United States,* 360 U.S. 109 (1959) 118; *Baggett* v. *Bullitt,* 377 U.S. 360 (1964) 369.

72 *Keyishian et al.* v. *Board of Regents of the University of the State of New York et. al.,* 385 U.S. 589 (1967), at p. 603. Quoted with approval in *University of Pennsylvania v. EEOC,* 493 U.S. 182 (1990), at p. 197; *Board of Education v. Pico,* 457 U.S. 853 (1982), at p. 870; *Regents of the Univ. of California v. Bakke,* 438 U.S. 265 (1978), at p. 312; *Epperson v. Arkansas,* 393 U.S. 97 (1968), at p. 105.

> "Academic Freedom, though not a specifically enumerated constitutional right, long has been viewed as a special concern of the First Amendment (…) Justice Frankfurter summarized the 'four essential freedoms' that constitute academic freedom: 'It is the business of the university to provide that atmosphere which is most conducive to speculation, experiment and creation. (…) to determine for itself on academic ground who may teach, what may be taught, how it shall be taught, and how may be admitted to study."[73]

That jurisprudence applies until today, though there has been criticism of those who maintained that it would have been unnecessary in the cases at bar to recognise a specific right of academic freedom given the guarantee of the freedom of speech[74]. Extensions of academic freedom as propounded by private associations such as the American Association of University Professors have not matured into legal rules[75]. Finally, the limits of academic freedom are discussed[76]. An annex to the freedom of science is the right to intellectual property, which – much to the astonishment of the European reader – has been included into the U.S. Constitution from the very beginning (Art. I, Sec. 8 § 8):

> "The Congress shall have Power … To promote the Progress of Science and useful Arts, by securing for limited Times to Authors and Inventors the exclusive Right to their respective Writings and Discoveries;…"[77]

---

73 *Regents of the University of California* v. *Bakke*, 438 U.S. 265 (1978), at p. 312.

74 Walter Metzger, "Profession and Constitution: Two Definitions of Academic Freedom in America", *Texas Law Review* 66 (1988), pp. 1265-1321; J. Peter Byrne, "Academic Freedom. A Special Concern of the First Amendment", *Yale Law Review* 99 (1989), pp. 251-339 at pp. 251 *et seqq.*

75 Cf. J. Peter Byrne, "Constitutional Academic Freedom after Grutter. Getting Real About the 'Four Freedoms' Of A University", *University of Colorado Law Review* 77 (2006), pp. 929-953 at p. 931.

76 Cf. Steven Goldberg, "The Constitutional Status of American Science", *University of Illinois Law Forum* 1 (1979), pp. 1-6 at pp. 1 *et seqq.*; US Congress – Office of Technology Assessment (ed.), *Science, Technology, and the First Amendment*, 1988, p. 37.

77 On this constitutional provision cf. Edward C. Walterscheid, "To Promote the Progress of Science and Useful Arts: The anatomy of a Congressional Power", *The Journal of Law and Technology* 43 (2002), pp. 1-81; id., "The Preambular Argument: the Dubious Premise of Eldred v. Ashcroft", *The Journal of Law and Technology* 44 (2004), pp. 331- 379 at pp. 339 *et seqq.*

Other constitutional jurisdictions follow the American way of guaranteeing. The Constitution of South Africa (1997) explicitly provides

> "Everyone has the right to freedom of expression, which includes ... academic freedom and freedom of scientific research."[78]

This strong guarantee is, of course, an answer to the oppressive regime that preceded today's constitutional situation in South Africa, but it is also a reflection of how freedom of science is considered following the American perspective of its proximity to freedom of speech. There has been some discussion on the precise content of freedom of science, but the core constitutional guarantee is not called into question[79].

Other countries in the American tradition – as far as academic freedom is concerned – are Indonesia[80], Egypt[81] and Nigeria[82]. India shares the same starting point[83], but extends the guarantee in some respect[84]. In

---

[78] Section 16 Constitution of South Africa (1997), available at http://www.constitutionalcourt.org.za/site/theconstitution/english-09.pdf.

[79] Cf. only John Higgins, "Academic Freedom in the New South Africa", in: *Boundary 2: International Journal of Literature and Culture* 27 (2000), pp. 97-119.

[80] Human Rights Watch (ed.), *Academic Freedom in Indonesia, Dismantling Soeharto-Era Barriers,* 1998 (http://www.unhcr.org/refworld/docid/3ae6a83a4.html).

[81] Articles 47 ("*Freedom of opinion shall be guaranteed. Every individual shall have the right to express his opinion and to publicise it verbally, in writing, by photography or by other means of expression within the limits of the law. Self criticism and constructive criticism shall guarantee the safety of the national structure.*") and 49 ("*The State shall guarantee for citizens the freedom of scientific research and literary, artistic and cultural creativity and provide the necessary means for its realization.*") of the Egyptian Constitution (1971, last amendment in 2007). It is however to be stated that "Red Lines" are drawn to exclude politics and religion from free research: Human Rights Watch (ed.), *Reading Between the "Red Lines". The Repression of Academic Freedom in Egyptian Universities,* 2005, at pp. 42 *et seq.*

[82] Articles 38 (1), 39 (1) and 40 of the Nigerian Constitution (1999). On the remarkable scientific efforts in that country cf. Babajide Alo, "University-Based Applied Research and Innovation in Nigeria", in: Osita Ogbu/Banji Oyeyinka/Hasa Mlawa, *Technology Policy and Practice in Africa,* 1995, http://www.idrc.ca/en/ev-30803-201-1-DO_TOPIC.html.

[83] Article 19 of the Indian Constitution (1950). See also the aspect of a duty in Article 51A of the Indian Constitution: "*It shall be the duty of every citizen of India ... (h) to develop the scientific temper, humanism and the spirit of inquiry and reform.*"

Brazil, freedom of science is also considered as a matter of free expression[85], and in the conflict with research interests this approach has shown some weakness in weighing freedom of science with other fundamental interests[86]. A similar situation can be found in Argentina[87], with an additional emphasis on the institutional academic independence of universities[88].

d) Absence of Constitutional Protection

It cannot be denied that there are constitutional jurisdictions devoid of any guarantee of freedom of science. Australia does not go beyond a

---

[84] The judgment of the Indian Supreme Court in *P. M. Bhargava & Ors.* v. *University Grants Commission & Anr.* (2002), available at http://www.indiankanoon.org/doc/697794/, is most illustrative: "Science is defined as knowledge acquired through the use of the scientific methods and the attributes of such knowledge include fallibility, verifiability and repeatability. Scientific truths are not dependant on whims and fancies of individuals. That apart science is international and if and when differences of opinion arise, scientists all over the world work honestly and diligently to resolve them.", and also the decision in *University of Mysore* v. *Govinda Rao AIR* 1965 SC 491: "The courts are not expert in academic matters and it is not for them to decide as what course should be taught in university and what should be their curriculum."

[85] Article 9 Brazilian Constitution 1988): "*... the expression of intellectual, artistic, scientific and communications activities is free, without any censorship or licence.*"

[86] Cf. Supremo Tribunal Federal, ADI 3.510, Rel. Min. Carlos Britto, julgamento em 28 e 29-5-08, Informativo 508, http://www.stf.gov.br//arquivo/informativo/documento/informativo508.htm, on the prohibition of research with human stem cells.

[87] Article 19 Argentinian Constitution (1853, last amendment in 1994): "*All the inhabitants of the Nation are entitled to the following rights ... to publish their ideas through the press without previous censorship ... to associate for useful purposes ... to teach and to learn.*"

[88] *Declaration of* unconstitutionality U. 2. XXXIII *of the Corte Suprema de Justicia de la Nación* Argentina, 27 May 1999, "Universidad Nacional de Córdoba c/ Estado Nacional"; decision N. 16. XXXIV of the Corte Suprema de Justicia de la Nación, 30 June 1998, "Niveyro, Silvia Mercedes c/ Universidad de Buenos Aires – resols. 2314/95"; decision M. 399 XXXIV of the Corte Suprema de Justicia de la Nación, 30 June 1998, "Monges, Analía M. c/ Universidad de Buenos Aires – resols. 2314/95". Cf. Ines Izaguirre, "Argentina", *Academe* Vol. 85 (1999), p. 18-20.

parliamentary report stating that even without an explicit constitutional safeguard, freedom of science is not at peril[89] and some scholarly utterances about the necessity of academic freedom as a matter of course[90]. The protection of free scientific activity in despotic regimes or least developed countries is far worse, however[91], and it is of course not viable to compare the Australian case with these instances.

3. Common Constitutional Elements

The overall picture on the constitutions of the various jurisdictions analysed is heterogeneous. As in the field of international instruments, there is no general rule to be "distilled" from a survey of domestic constitutions of the world, but parallel to the analysis of international principles and rules, similar elements appear to be obvious to sound legal regulation of science. First of all, it is the free exchange of scientific reflections and their results which can be considered a common denominator of all jurisdictions – the obvious dictatorial incursions into that right put aside. Secondly, it is the awareness of the social (and not just economic) need to bring forward science in whatever jurisdiction. Thirdly, it is the recognition of some institutional structures to enable and facilitate science to be performed. An annex may be the right of the scientist to be identified (also economically) as the owner of his very thoughts and achievements.

IV. A Constitutional Point of Orientation

Albeit sometimes in rudiments only, there are indications for the recognition of a right to free science at the international, regional and domes-

---

[89] The Senate. Standing Committee on Education, Employment and Workplace Relations (ed.), *Report Allegations of Academic Bias in Universities and Schools,* 2008.

[90] Michael Herriman, "Academic Freedom in Australia", *Interchange* 14 (1983), pp. 82-93; Carol Kayrooz/Pamela Kinnear/Paul Preston, "Academic Freedom and Commercialisation of Australian Universities: Perceptions and Experiences of Social Scientists", *Australia Institute Discussion Paper* No.37 (2001), p. 44.

[91] Sebastian Steinecke (supra note 13), manuscript pp. 260 *et seq.*

tic level. Some elements at least are beyond doubt: The freedom to exchange ideas and to publish in scientific matters, marginally annexed by the right to intellectual property, the recognition of the benefits of free science and the concomitant necessity of providing support and basic organisational facilities for science. These elements also indicate that the freedom of scientific research has several dimensions: In some instances, protection against intrusions from national or international authorities is required, whereas in other instances, it is these authorities, who have to become active for the preservation of science. The ambivalent relationship between science and regulatory forces (see above A. IV.) reappears. It is finally by no means denied that free science can be subject to limits.

A terminological point should be added. Some instruments – whether at the global, regional or domestic level – use the term "academic freedom" instead of a terminology applying the expressions science or research. There are however no indications at any level whatsoever that more lies behind this terminological discrepancy but an allusion to traditional developments. It is true that for a long time, scientific research was confined to universities, academies and similar institutions, but it cannot be denied that throughout the various industrial revolutions, it has been considerably extended e.g. to commercial bodies. The limits of "science" in such bodies in terms of independence have already been touched upon (above A. II. 2.), but there is no need to exclude them from the freedom of science as such. Therefore, we submit that there is no difference between freedom of science and academic freedom, and this non-differentiation directly hints to the institutional plurality of the governance of science.

D. Institutional Design

I. Global Administrative Law and Institutional Thinking

Within the idea of global administrative law, institutions are playing a major role. While it is certainly not helpful to confine global administrative law to global administration in the sense of a limitation to institutional matters, it is indispensable to take a closer look on the existing institutional framework, its problems and its potential. This analysis takes the traditional public international legal perspective as a starting point, where states are considered as "institutions" in the first place (below II.). It further follows this line by considering International Organisations including the European Union as the only supranational institution in the field (below III.), before turning to the "novelty" of global governance and global administrative law, the network structures encompassing sub-state entities as well as private bodies (below IV.).

II. States

Among the actors in the international governance of science, States have to be mentioned in the first place. This may appear astonishing, if not confusing, as States are responsible above all for domestic regulation and government and do not possess unilateral powers for international rulemaking and execution. However, while considering the broad interest and even enthusiasm for new structures of governance, it should not be forgotten that States continue to be the main subjects of international law[1] and that they are the only subjects that are conferred personality by their mere existence[2]. Beyond these legal reflections, it can-

---

1 Cf. the seminal work of Peter Saladin, *Wozu noch Staaten?: zu den Funktionen eines modernen demokratischen Rechtsstaats in einer zunehmend überstaatlichen Welt,* 1995.

2 Ian Brownlie, *Principles of Public International Law*, 7th ed. 2008, p. 57.

not be denied that States are considerably active in the field of science in their own economic interest[3].

Thus, States are developing broad activities in promoting science. One possible way to effectuate this is the operation or funding of research programmes and projects by government ministries. It is common, to give an example, that German Federal Ministries undertake research efforts either in internal sub-units, by issuing research contracts to independent researchers or by creating research agencies under governmental auspices[4]. Other national research agents are well known at global level such as NASA in the U.S.[5] or MITI in Japan[6]. Academies of science[7] can have such a proximity to government that they cannot duly be granted the benefit of freedom of science: It is most questionable that a government institution, i.e. the emanation of the relevant State, should be vested with elements of a human right to be directed against government institutions itself.[8] It should finally be noted that networks of sub-state-entities are considered separately (below IV.).

---

3 This is historically founded since the 1870s: Dominique Pestre, "The Technosciences between Markets, Social Worries and the Political: How to Imagine a Better Future", in: Helga Nowotny/Dominique Pestre/Eberhard Schmidt-Aßmann/Helmuth Schulze-Fielitz/Hans-Heinrich Trute, *The Public Nature of Science under Assault*, 2005, pp. 29-52 at p. 32.

4 Cf. also the "Initiative Außenwissenschaft" to promote German Scientific Activity abroad by the Auswärtiges Amt (http://www.auswaertiges-amt.de/diplo/de/Infoservice/Presse/Meldungen/2009/090312-AWP.html).

5 On NASA see http://www.nasa.gov/; Committee on NASA's Suborbital Research Capabilities/National Research Council, *Revitalizing NASA's Suborbital Program: Advancing Science, Driving Innovation, and Developing a Workforce*, 2010; Committee for the Review of NASA's Revolutionize Aviation Program, Aeronautics and Space Engineering Board, Division on Engineering and Physical Sciences/National Research Council of the National Academies, *Review of NASA's aerospace technology enterprise: an assessment of NASA's Pioneering Revolutionary Technology Program*, 2004.

6 On MITI see Thomas Neuschwander, *Mythos MITI: Industriepolitik in Japan*, 1994; Scott Callon, *Divided sun: MITI and the breakdown of Japanese high-tech industrial policy, 1975-1993*, 1995.

7 But see below IV. 2. b.

8 This is most controversial in German constitutional law; see on the one hand Matthias Ruffert, "Grund und Grenzen der Wissenschaftsfreiheit", *Veröffentlichungen der Vereinigung der Deutschen Staatsrechtslehrer* 65 (2006), pp. 146-210 at pp. 178 *et seq.*, on the other hand Christian Starck, in: Hermann von

Furthermore, the domestic regulatory power of States causes effects in the international sphere, in particular with respect to national rules protecting a certain good, right or value. The greater the rigidity of such rules, the more significant the potential risk for a State that research activity would be undertaken elsewhere. From an isolated perspective of freedom of science, this might be considered as an advantageous effect lying at the heart of many processes agglutinated in the concept of globalisation[9]. The *Pasteur*-quotation can be repeated here: "*Le savant a une patrie, la science n'en a pas*". Science searches the place (i.e. jurisdiction) of optimum guarantee (i.e. protection against government interference)[10]. It is generally beyond doubt that competitive structures amongst States are able to promote the benefits of free activities. What is also beyond doubt is the potential danger of a "race-to-the-bottom". Research that is perilous to the environment or that impedes upon commonly accepted ethical values may be transferred to States with a low regulatory level and – as it is to be feared – with deficits of control. If consensus can be established to avoid such deprecation of values, the need for the use of international governance tools is underlined[11].

Mangoldt/Friedrich Klein/id. (eds.), *Kommentar zum Grundgesetz,* Vol. 1, 6th ed. 2010, Article 5, para. 409.

[9] Martin Nettesheim, "Grund und Grenzen der Wissenschaftsfreiheit", *Deutsches Verwaltungsblatt* 17 (2005), pp. 1072-1081 at p. 1073. Generally Lüder Gerken, *Der Wettbewerb der Staaten,* 1999; Karl M. Meessen, "Souveränität im Wettbewerb der Systeme", in: Volkmar Götz/Peter Selmer/Rüdiger Wolfrum (eds.), *Liber amicorum Günther Jaenicke – Zum 85. Geburtstag*, 1998, pp. 667-681; Jean-Bernard Auby, *La globalisation, le droit et l'État,* 2003, pp. 86 *et seq.*, Matthias Ruffert, *Die Globalisierung als Herausforderung an das Öffentliche Recht,* 2004, pp. 43 *et seq.*

[10] See Volker Röben, "The Sciences – A Contribution to Understanding the Law on an Activity of International Concern", German Yearbook of International Law 37 (1994), pp. 254-280 at p. 278.

[11] See Stefan Kadelbach, "Demokratische Legitimation als Prinzip zwischenstaatlichen Handelns", in: Silja Vöneky/Cornelia Hagedorn/Miriam Clados/Jelena von Achenbach (eds.), *Legitimation ethischer Entscheidungen im Recht,* 2009, pp. 147-171 at p. 148.

III. International Organisations

1. Universal Organisations and their Activity in Science

a) Basic Notions

The strong position of International Organisations in international governance is generally beyond any doubt. The rise of such organisations since 1945 has often been described, and although the empirics of the rise are not easy to be displayed in exact figures, there is hardly any controversy about their overwhelming importance in the current age of international law[12]. Insecure empirical data may not be an obstacle to a sound legal analysis, but viable juridical definitions are, on the contrary, indispensable.

In this context, International Organisations shall be defined as organisational structures aggregating subjects of international law (above all States) on the basis of an international treaty, accomplishing tasks of common interest.[13] This definition excludes first of all associations of non-State actors such as non-governmental organisations (NGOs) which shall be treated separately (below IV., V.). It further does not comprise structures and regimes based on legally non-binding agreements which also require particular attention in other contexts (below IV.). To categorise the institutional actors, it is useful to consider that a typology of universal and regional organisations finds overall acceptance[14]. Universal organisations are active at global level, whereas regional organisations are confined to a certain number of States. This confinement is generally performed on a geographical basis, but it can also consider economical and political aspects such as in the OSCE (Europe plus the U.S. and Canada) or the OECD (Europe plus other States of the world with "Western" economies). Finally, the category of supranational organisations will be dealt with in the context of the EU (below 2.).

---

[12] For the recent history of International Organisations see Christian Walter and Matthias Ruffert, *Institutionalisiertes Völkerrecht,* 2009, paras. 53 *et seq.*; Volker Rittberger and Bernhard Zangl, *Internationale Organisationen*, 3rd ed. 2003, pp. 49 *et seqq.;* Guiseppe Schiavone, *International Organizations – A Dictionary and Directory,* 7th edition 2008.

[13] Definition taken from Christian Walter and Matthias Ruffert (supra note 12), para. 9.

[14] See Christian Walter and Matthias Ruffert (supra note 12), para. 13.

b) UNESCO

Among the international organisations acting at universal level, the United Nations Educational, Scientific and Cultural Organisation (UNESCO) is certainly the most prominent one, and it is appropriate to deal with it first for several reasons. In general, UNESCO is an organisation with a typical design: Founded upon an international treaty (the UNESCO-Constitution of 1945[15]), made up by basically three main organs (General Conference, Executive Board and Secretariat)[16] and linked to the "UN-family" by a treaty of co-operation based on Articles 57 and 63 of the UN Charter[17]. In particular, UNESCO is the only universal international organisation bearing the term "Science" in its official denomination.

This is also reflected in the Constitution of UNESCO which designates the promotion of international cooperation in the field of science as an important aim of the institutions, be it generally, through the conservation of monuments of scientific history or by means of the promotion of the international mobility of scientists (Article I (1) and I (2) (c)).

> "*Article I: Purposes and functions*
>
> 1. The purpose of the Organization is to contribute to peace and security by promoting collaboration among the nations through education, science and culture in order to further universal respect for justice, for the rule of law and for the human rights and fundamental freedoms which are affirmed for the peoples of the world, without distinction of race, sex, language or religion, by the Charter of the United Nations.
>
> 2. To realize this purpose the Organization will:
>
> ...
>
> (c) Maintain, increase and diffuse knowledge:
>
> ...
>
> By encouraging cooperation among the nations in all branches of intellectual activity, including the international exchange of persons active in the fields of education, science and culture and the ex-

---

15 4 U.N.T.S. 275.

16 Philippe Sands and Pierre Klein, *Bowett's Law of International Institutions*, 6th ed. 2009, paras. 3-027 to 3-030.

17 1 U.N.T.S. 238. (twice amended 1948 and 1962).

change of publications, objects of artistic and scientific interest and other materials of information; ..."

Within the organisation, two of the five specialised sectors are directly related to scientific activity, i.e. Natural Sciences and Social and Human Sciences. The activities of the organisation lie mainly in the organisation of international thematic conferences[18].

In terms of legislation, the General Conference is empowered to issue recommendations which are not strictly binding on the Member States but which can trigger reporting requirements (Articles IV (B) (4) and VIII of the Constitution). The General Conference can also submit proposals for the conclusion of international treaties (Article IV (B) (4)), and it can issue Declarations.

> "4. The General Conference shall, in adopting proposals for submission to the Member States, distinguish between recommendations and international conventions submitted for their approval. In the former case a majority vote shall suffice; in the latter case a two-thirds majority shall be required. Each of the Member States shall submit recommendations or conventions to its competent authorities within a period of one year from the close of the session of the General Conference at which they were adopted."

This has been done extensively[19].

Among the strategic activities of UNESCO, the creation of subsidiary bodies, organs or institutionalised programmes plays an important role. In this context the organisation has set up the World Commission on the Ethics of Scientific Knowledge and Technology (COMEST) in 1998 as an advisory body composed of 18 independent experts[20]. COMEST shall mainly formulate ethical principles to give advice to decision-makers. On a different level, UNESCO hosts the International Council for Philosophy and Humanistic Studies (ICPHS), a non-governmental

---

[18] On UNESCO see: Klaus Hüfner and Wolfgang Reuther (eds.), *UNESCO-Handbuch*, 2nd ed. 2005; Foreign Office of the Federal Republic of Germany and German Commission for UNESCO, *The Role of UNESCO. A Contribution to the Debate by Germany*, 2007.

[19] For an overview see: http://portal.unesco.org/en/ev.php-URL_ID=23772&URL_DO=DO_TOPIC&URL_SECTION=201.html.

[20] For its constitution see: http://www.unesco.org/new/en/social-and-human-sciences/themes/ethics-of-science-and-technology/science-and-technology/comest/.

organisation with its own institutional framework[21]. ICPHS comprises a considerable number of scientific societies in the fields of philosophy, human sciences and related fields. Finally, there is the programme "Management of Social Transformations" (MOST) to promote social sciences and to guarantee certain ethical standards with these sciences.[22]

c) Other Organisations within the Framework of the UN

aa) International Atomic Energy Agency (IAEA)

Despite its particular status as an agency that is not a specialised agency under Articles 57 and 63 of the UN-Charter (the respective agreement was concluded with the General Assembly and not with the Economic and Social Committee)[23], despite the great impact of questions of international peace and security and despite the controversy on the use of nuclear energy, the International Atomic Energy Agency (IAEA) is, next to UNESCO, the organisation within the UN-framework with the most considerable activity and weight in research matters. There is a clear legal basis for this orientation towards research in Article III (A) of the IAEA's Statute (1957)[24], as the Agency is empowered to encourage and to assist research on the peaceful use of atomic energy, including the provision of means and the encouragement of the exchange of experts and knowledge.

> *Article III:* Functions
>
> A. The Agency is authorized:
>
> 1. To encourage and assist research on, and development and practical application of, atomic energy for peaceful uses throughout the world; and, if requested to do so, to act as an intermediary for the purposes of securing the performance of services or the supplying of materials, equipment, or facilities by one member of the Agency for another; and to perform any operation or service useful in research on, or development or practical application of, atomic energy for peaceful purposes;

---

[21] As for its constitution cf. http://portal.unesco.org/culture/en/ev.php-URL_ID=1590&URL_DO=DO_TOPIC&URL_SECTION=201.html.

[22] Cf. http://www.unesco.org/new/en/social-and-human-sciences/themes/social-transformations/most-programme/.

[23] Cf. Philippe Sands and Pierre Klein (supra note 16), para. 3-075.

[24] 276 U.N.T.S. 3.

2. To make provision, in accordance with this Statute, for materials, services, equipment, and facilities to meet the needs of research on, and development and practical application of, atomic energy for peaceful purposes, including the production of electric power, with due consideration for the needs of the under-developed areas of the world;

3. To foster the exchange of scientific and technical information on peaceful uses of atomic energy;

4. To encourage the exchange of training of scientists and experts in the field of peaceful uses of atomic energy; ..."

Consequently, the organisation aims in its current Medium Term Strategy (2006-2011) at the advancement of nuclear science and technology including information on it[25].

To implement Article III of its statute, the IAEA undertakes operative research projects, the so-called Coordinated Research Projects (CRPs), bringing together research institutes from the Member States. The publication of the research results is also managed by the agency[26].

bb) Commission on Sustainable Development (CSD)

In the field of environmental protection, the Commission on Sustainable Development (CSD)[27] has been established following up the UN Conference on Environment and Development in Rio 1992[28]. Formally, it is one of the special commissions of the ECOSOC. Amongst many other functions, the CSD also undertakes research activities[29] in implementing the decisions of UNCED including the broad action programme Agenda 21. The programme itself has comprehensive chapters on science and technology (Chapter 1) and on science serving sustain-

---

[25] Available at: http://www.iaea.org/About/mts2006_2011.pdf .

[26] See http://www-crp.iaea.org/default.asp.

[27] GA Res. 47/191, ECOSOC Res. 1993/207. See also http://www.un.org/esa/sustdev/csd/csd12/csd12.htm.

[28] On UNCED see Matthias Ruffert, "Das Umweltvölkerrecht im Spiegel der Erklärung von Rio und der Agenda 21", *Zeitschrift für Umweltrecht* 1993, pp. 208-214.

[29] See also the Report http://www.bmu.de/files/pdfs/allgemein/application/pdf/csd_01.pdf at pp. 99 and 106-109.

able development (Chapter 35)[30]. The later Summit in Johannesburg passed a Plan of Implementation of the World Summit on Sustainable Development including the aim to enhance research for sustainable development[31].

cc) Food and Agricultural Organisation (FAO)

Finally, the activities of the Food and Agricultural Organisation (FAO) should be mentioned, which has established a Commission on Genetic Resources for Food and Agriculture (CGRFA) to deal with respective questions of research[32].

d) Organisation for Economic Co-Operation and Development (OECD)

A leading textbook on the law of international organisations notes that: "An assessment of the work of OECD, and even a proper description of its activities, is probably more a task for the economist than for the lawyer."[33] In many fields – including research – the OECD appears to be one of the most underestimated international institutions as far as its actual policy performance is concerned. This performance is disproportionate to the legal studies about the organisation – the leading work in German dates from the 1970s, and the literature about the organisation in English is even older[34].

---

30 Preamble (Chapter 1): http://www.un.org/esa/dsd/agenda21/res_agenda21_01.shtml; Chapter 35: http://www.un.org/esa/dsd/agenda21/res_agenda21_35.shtml. Cf. also Principle 9 of the Rio Declaration on Environment and Development, UN Doc. A/CONF.151/26 (Vol. I).

31 Available at: http://www.un.org/esa/sustdev/documents/WSSD_POI_PD/English/WSSD_PlanImpl.pdf. See in particular pages 19-20, 28, 36, 42, 50-51 and 59.

32 http://www.fao.org/nr/cgrfa/en/.

33 Philippe Sands and Pierre Klein (supra note 16), para. 6-054.

34 Hugo J. Hahn and Albrecht Weber, *Die OECD: Organisation für Wirtschaftliche Zusammenarbeit und Entwicklung*, 1976; Henry G. Aubrey, *Atlantic Economic Cooperation: The Case of the OECD*, 1967; Hugo J. Hahn, "Organisation for Economic Co-operation and Development", in: Rudolf Bernhardt (ed.), *Encyclopedia of Public International Law*, Vol. 3 (1997), pp. 790-799, and the descriptive entry in Guiseppe Schiavone (supra note 12), pp. 262-266. But see the specific study in social sciences by Kerstin Marten and

The OECD is a classical international organisation created by a treaty in 1960[35]. A particularity is that it succeeds the Organisation for European Economic Co-Operation, which had been established in 1948 to organise the distribution of Marshal Plan funds in post-war (Western) Europe[36]. According to Article 1 of its Convention, the OECD basically aims at the support of economic growth and development within its Member States, which are mainly European, but also comprise the US and Canada and other industrialised countries outside Europe. In the field of science and technology, it is agreed in Article 2 of the Convention that

> *Article 2*
>
> In the pursuit of these aims, the Members agree that they will, both individually and jointly: ...
>
> (b) ...promote the development of their resources, encourage research and promote vocational training; ...".

If we consider the broad activities of OECD in research policy, it is most astonishing that they are developed on such a short legal basis.

What is more, OECD's formal powers to legislate are rather limited. Under Article 5 of the Convention, the organisation is entitled to take binding decisions, make recommendations or enter into international agreements (with Members, non-members and international organisations).

> "Article 5
>
> In order to achieve its aims, the Organisation may:
>
> (a) take decisions which, except as otherwise provided, shall be binding on all the Members;
>
> (b) make recommendations to Members; and
>
> (c) enter into agreements with Members, non-member States and international organisations."

However, decision-making needs unanimity, and according to Article 6 (3), first sentence of the Convention

> "No decision shall be binding on any Member until it has complied with the requirements of its own constitutional procedures."

Klaus Dieter Wolf, "Paradoxien der Neuen Staatsräson", *Zeitschrift für Internationale Beziehungen,* 13 (2006), pp. 145-176.

[35] 888 U.N.T.S. 179.

[36] 888 U.N.T.S. 141.

Therefore, it is most unusual that OECD pursues its policies by the use of binding decisions. Sometimes, guidelines are passed, but what is far more "popular" in the context of that organisation is governance by means of reports containing benchmarks and descriptions of Member States' performance[37].

2. Supranational Institutions of Research: The European Union

a) The EU as a Supranational Organisation in the International Field of Research

It is beyond doubt that the EU is a strong actor in the field of research – as in many policy fields. It might however be doubted whether the EU could be included into the framework of international law, as it represents a very high level of integration of the participating States that transcends ordinary inter-state cooperation. These doubts would duly reflect the process of separation between EU law and international law caused by the emergence of EU law as a particular category. They are however unjustified if they encompass the idea that supranational institutions are an institutional pattern to be strictly separated from international organisations[38]. On the contrary, supranationality is supposed to be a distinctive quality of international organisations, but nothing precludes the observer from having a look at supranational organisations from the international (institutional) lawyer's perspective. The existence of a constitutional framework of European integration does not exclude it from international law approaches.

The Treaty on the Functioning of the European Union (TFEU) as designed by the Lisbon Reform Treaty[39], contains a line of articles on research, technological development and space, in continuation of the relevant treaty section that was established in the Maastricht Treaty in

---

[37] Hugo J. Hahn, "Organisation for Economic Co-operation and Development", in: Rudolf Bernhardt (ed.), *Encyclopedia of Public International Law*, Vol. 3 (1997), pp. 790-799 at p. 797.

[38] On this problem cf. Christian Walter and Matthias Ruffert (supra note 12), paras. 16-17.

[39] Treaty of Lisbon amending the Treaty on European Union and the Treaty establishing the European Community, signed at Lisbon, 13 December 2007, [2007] O.J. C 306/1.

1993[40]. The main novelty at institutional level is the integration of the "European Research Area" into the central treaty provision (Article 179 TFEU):

> Article 179 (ex Article 163 TEC)
>
> 1. The Union shall have the objective of strengthening its scientific and technological bases by achieving a European research area in which researchers, scientific knowledge and technology circulate freely, and encouraging it to become more competitive, including in its industry, while promoting all the research activities deemed necessary by virtue of other Chapters of the Treaties.

Furthermore, there is now a particular article (Article 189 TFEU) on European space policy which integrates the European Space Agency (ESA) on a co-operative basis (Article 189 (3) TFEU)[41].

The relevant powers are shared competences according to Art. 2 (2) TFEU, but in a limited way: The general rule in Article 2 (2) TFEU provides that in fields of shared competence the Union and the Member States may legislate and adopt legally binding acts in that area, but that the Member States shall exercise their competence only to the extent that the Union itself has not exercised its competence. Article 4 (2) TFEU does not include research, technological development and space in the list of areas where the general rule of shared competences is applicable, but contains a particular section (3), which provides:

---

40 Cf. Christine Godt, "Forschungs-, Wissenschafts- und Technologiepolitik", in: Manfred Dauses, *Handbuch des EU-Wirtschaftsrechts,* 2009, Chapter N, pp. 1-28; Matthias Ruffert, in: Christian Calliess/id., (eds.), *EUV/AEUV Kommentar*, 4th ed. 2010, Art. 179, para. 5; Henning Eikenberg, in: Eberhard Grabitz and Meinhard Hilf (eds.), *Das Recht der Europäischen Union, Band I EUV/EGV*, 2005, Preliminaries to Art.163-173, para. 27; Overview of the development of the European research policy: Juliane Hilf, in: Hans von der Groeben and Jürgen Schwarze (eds.), *Kommentar zum Vertrag über die Europäische Union und zur Gründung der Europäischen Gemeinschaft*, 6th ed. 2003, Preliminaries to Art. 163-173, paras. 53 *et seqq.*

41 On this co-operation see Annette Froehlich, "Die European Space Policy und ihre Auswirkungen auf die Verlängerung des ESA/EG-Rahmenabkommens", *Zeitschrift für Luft- und Weltraumrecht* 57 (2008), pp. 67-77; Stephan Hobe/Katharina Kunzmann/Thomas Reuter/Julia Neumann, *Rechtliche Rahmenbedingungen einer künftigen kohärenten Struktur der Europäischen Raumfahrt*, 2005; Thomas Reuter, *Die ESA als Raumfahrtagentur der Europäischen Union*, 2007.

> "In the areas of research, technological development and space, the Union shall have competence to carry out activities, in particular to define and implement programmes; however, the exercise of that competence shall not result in Member States being prevented from exercising theirs."

EU-Governance in the field of research is thus mainly supposed to be exercised by EU programmes without preventing the Member States from exercising their powers[42]. This is reflected in the design of Articles 179 to 190, which put programmes in the centre of EU activity (Article 180 lit. a, 182 TFEU) together with measures implementing the European Research Area[43].

b) The European Research Area

The establishment of the European research area lies at the heart of the EU's research policy. According to Article 179 (1) TFEU, its creation is the main goal of the Union in the field of research, technological development and space. That area shall be characterised by the free circulation of researchers, scientific knowledge and technology[44].

The development of the idea of a European research area together with the additional wording of Article 179 (1) TFEU clearly shows the direction of the Union's research policy. From the outset, the area was designed to enhance competitiveness and employment in the EU, and it is part of the Lisbon strategy in economic policy that aimed at making Europe the most dynamic, innovative and competitive market from 2000 to 2010[45]. Therefore, it shall be the aim of the European research

---

42 This is underlined by Declaration No. 24 to the TFEU (Declarations annexed to the Final Act of the Intergovernmental Conference which adopted the Treaty of Lisbon, [2008] O.J. C 115/335), on which cf. Josef Franz Lindner, *Die Europäisierung des Wissenschaftsrechts,* 2009, p. 49.

43 Cf. Matthias Ruffert (supra note 40), Art. 179, paras. 8 *et seqq.*

44 Cf. Article 179 (1) TFEU; cf. Matthias Ruffert (supra note 40), Art. 179, para. 8; Juliane Hilf (supra note 40), Preliminaries to Art. 163-173, para. 73; Henning Eikenberg (supra note 40), Preliminaries to Art.163-173 (2005), paras. 30 *et seqq.*

45 Cf. Matthias Ruffert (supra note 40), Art. 173, para. 7; Christine Mellein, in: Jürgen Schwarze (ed.), *EU-Kommentar*, 2nd ed. 2009, Art. 163, para. 9. Although the strategy was far from being successful, there shall be a continuation until 2020: *European Commission,* Consultation on the future "EU 2020" strategy, Doc. COM(2009) 647 final; Communication from the Commission

area to strengthen the Union's scientific and technological bases, and a core objective accompanying the creation of the area is the encouragement of the competitiveness of the Union "including in its industry"[46]. The 2007 Green Paper (The European Research Area: New Perspectives)[47] lays down five main objectives:

- adequate flow of competent researchers,
- world-class research infrastructures,
- excellent research institutions engaged in effective public-private cooperation,
- effective knowledge-sharing notably between public research and industry, as well as with the public at large,
- well-coordinated research programmes and priorities.

The issuing of the Green Paper was followed by a broad consultation process in 2007 in which many relevant stakeholders participated[48], and it lead to the reformulation of the goals of the area up to the year 2020[49]. To summarise, the European research area focuses upon economic progress and growth through support in research. We shall turn to the institutional framework established in this context and beyond.

c) *The Institutional Framework*

aa) The Commission

The EU's research policy is embedded in the ordinary institutional framework of the Union. The triangular combination of European Par-

---

"EUROPE 2020 – A strategy for smart, sustainable and inclusive growth", COM(2010)2020 final and European Council conclusions of 17 July 2010, Doc. EUCO 13/10.

[46] Cf. Article 179 (1) TFEU.

[47] Doc. COM(2007) 161 final.

[48] Cf. Commission, 2 April 2008, SEC(2008) 430: Results of the Public Consultation on the Green Paper "The European Research Area: New Perspectives"; Henning Eikenberg, "Der Europäische Forschungsraum: Ein Kompetenzproblem?", in: *Europarecht* 2008, pp. 125-139 at pp. 135 *et seq*.

[49] Cf. Council conclusions on the definition of a "2020 Vision for the European Research Area" (Competitiveness Council, 2 December 2008) – http://www.consilium.europa.eu/ueDocs/cms_Data/docs/pressData/en/intm/104434.pdf.

liament, Council of Ministers and European Commission[50] in that framework is designed to balance interests of the supranational union, the Member States and the citizens[51]. The basic programming powers are exerted according to the ordinary legislative procedure (Articles 182 (1) with Articles 289 and 294 TFEU). This does not mean, however, that the factual competences between the institutions are distributed on an equal basis. The position of the commission in drafting the programmes, in preparing new initiatives on the basis of green papers, hearings and consultations, is considerably strong. The "monopoly of initiative"[52] continues to be vested in the Commission, whose Directorate General for research is particularly active in that field, and it does not count less but 14 directorates![53] Research matters in the Commission alone are assigned € 50.5 bn, according to the Seventh Framework Programme, which represents 41,09 % of the EU budget (€ 122,9 bn in 2010)[54].

The Commission also undertakes own research activities. The Joint Research Centre (JRC) started from research in the context of Euratom, but now encompasses research in various fields of scientific research in five facilities[55]. The JRC is also involved in giving policy advice to the Commission in the field of research. Its legal institutional status is a purely internal one: it is one of the General Services of the Commission bearing the rank of a Directorate-General[56].

---

50 The ECJ could be mentioned in the context of applying scientific expert advice in particular cases: Joseph Corkin, "Science, Legitimacy and the Law: Regulating Risk Regulation Judiciously in the European Community", *European Law Review* 33 (2008), pp. 359-384.

51 Cf. only Derrick Wyatt/Alan Dashwood/Michael Dougan/Anthony Arnull/Malcolm Ross/Elanor Spaventa, *Wyatt & Dashwood's European Union Law,* 5th ed. 2006, Chapter 2; Damian Chalmers/Gareth Davies/Giorgio Monti, *European Union Law,* 2nd ed. 2010, pp. 52 *et seq.*

52 Damian Chalmers/Gareth Davies/Giorgio Monti (supra note 51), at p. 61.

53 Cf. Official Directory of the European Union, *European Communities,* 2009, pp. 260 *et seqq.*; Directorate-General for Research: http://ec.europa.eu/dgs/research/organisation.cfm?lang=en.

54 Cf. Matthias Ruffert (supra note 40), Art. 182, para. 4.

55 Cf. Juliane Hilf (supra note 40), Preliminaries to Art. 163-173, paras. 38 *et seqq.*

56 Cf. Matthias Ruffert (supra note 40), Art. 179, para. 17; Juliane Hilf (supra note 40), Preliminaries to Art. 163-173, para. 45.

bb) Organised Co-operation between Member States

Beyond central programming, but interlinked with Commission activity, the organised co-operation between Member States is an important field of EU research policy. Such co-operative structures are envisaged in many articles of the TFEU[57], and their result is the establishment of a governance network of shared management (*Verwaltungsverbund*)[58]. Thus, following Article 181 TFEU, the EU and the Member States "shall coordinate their research and technological development activities so as to ensure that national policies and Union policy are mutually consistent" which is promoted by the Commission using guidelines and indicators, the exchange of best practice and "elements for periodic monitoring and evaluation"[59]. This kind of co-ordination procedure, well-known as "open method of co-ordination" is a steering tool which appears to be "soft" in legal terms as benchmarking indicators and their evaluation do not create binding legal effects, but which can have enormous political consequences[60]. We shall deal with this mechanism in the context of implementation and management. Furthermore, ac-

---

[57] Art. 67 (3), Art. 74, Art. 76, Art. 82, Art. 87 (1) TFEU (operational cooperation between the authorities including police, customs and other investigative and law enforcement authorities (Police and Judicial Co-operation in Criminal Matters)); Art. 19 TFEU (the field of anti-discrimination legislation); Art. 46 (a) TFEU (*cooperation* between national *employment services*); Art. 50 (2) (b) TFEU (administrative cooperation in the field of the right of establishment); Art. 33 TFEU (customs); Art. 149, Art. 150 TFEU (administrative cooperation with regard to employment measures); Art. 153 (2) (a), Art. 156, Art. 160 TFEU (administrative cooperation in the field of social policy).

[58] On this concept see Eberhard Schmidt-Aßmann, in: id. and Bettina Schöndorf-Haubold (eds.), *Der Europäische Verwaltungsverbund*, 2010, pp. 1 *et seqq.*; Herwig C.H. Hofmann and Alexander Türk, "The Development of Integrated Administration in the EU and its Consequences", *European Law Journal* 13 (2007), pp. 253-271; further Gernot Sydow, "Die Vereinheitlichung des mitgliedstaatlichen Vollzugs des Europarechts in mehrstufigen Verwaltungsverfahren", *Die Verwaltung* 34 (2001), pp. 517-542 at pp. 520 *et seqq.*; Sabino Cassese, "European Administrative Proceedings", *Law and Contemporary Problems* 68 (2004), pp. 21-35; Wolfgang Weiß, *Der Europäische Verwaltungsverbund*, 2010; Matthias Ruffert, "Von der Europäisierung des Verwaltungsrechts zum Europäischen Verwaltungsverbund", *Die Öffentliche Verwaltung* 2007, pp. 761-770.

[59] Article 181 (2) TFEU.

[60] Cf. Matthias Ruffert (supra note 40), Art. 181, para. 3; Henning Eikenberg (supra note 40), Art. 165, paras. 17 *et seq.*

cording to Article 185 TFEU, the Union may co-operatively participate in research and development programmes undertaken by several Member States[61].

cc) Special Agencies

(1) European Research Council (ERC)

The creation of agencies is a trend in European institutional and administrative law that has been developing for about ten years[62]. Regulatory agencies are created to establish independent institutions which are particularly competent in a certain policy field[63]. Executive agencies are

---

[61] Cf. Matthias Ruffert (supra note 40), Art. 185; Henning Eikenberg (supra note 40), Art. 169; Juliane Hilf (supra note 40), Art. 169.

[62] An important step was the Commission White Paper on Governance in Europe: European governance – a White Paper, [2001] O.J. C 287/1 (pp. 19 *et seqq.*); Communication from the Commission – The operating framework for the European Regulatory Agencies, Doc. COM(2002) 718 final, further Draft Interinstitutional Agreement on the operating framework for the European regulatory agencies, Doc. COM(2005) 59 final.

[63] On the development of the EU regulatory agencies see the following publications: Michael Berger, *Vertraglich nicht vorgesehene Einrichtungen des Gemeinschaftsrechts mit eigener Rechtspersönlichkeit*, 1999, pp. 31 *et seqq.*; Paul Craig, *EU Administrative Law*, 2006, pp. 148 *et seqq.*; Edoardo Chiti, "The Emergence of a Community Administration: the Case of European Agencies", *Common Market Law Review* 37 (2000), pp. 309-334 at p. 309; id. "Towards a Model of Independent Exercise of Community Functions?", in: Roberto Caranta/Mads Andenæs/Duncan Fairgrieve (eds.), *Independent Administrative Authorities* (2004), pp. 205-224; Georg Hermes, "Legitimationsprobleme unabhängiger Behörden", in: Hartmut Bauer/Peter M. Huber/Karl-Peter Sommermann (eds.), *Demokratie in Europa*, 2005, pp. 457-490 at pp. 460 *et seqq.*; Arno Kahl, "Europäische Agenturen im Lichte der dynamischen Verwaltungslehre", in: Konrad Arnold/Friederike Bundschuh-Rieseneder/id./Thomas Müller/Klaus Wallnöfer (eds.), *Recht Politik Wirtschaft Dynamische Perspektiven – Festschrift für Norbert Wimmer*, 2008, pp. 245-284; Helmuth Schulze-Fielitz, in: Wilfried Erbguth and Johannes Masing (eds.), *Verwaltung unter dem Einfluß des Europarechts*, 2006, pp. 91-136 at p. 97; Ellen Vos, "Reforming the European Commission: What Role to Play for European Agencies", *Common Market Law Review* 37 (2000), pp. 1113-1134 at pp. 1116 *et seqq.*; Robert Uerpmann, "Mittelbare Gemeinschaftsverwaltung durch gemeinschaftsgeschaffene juristische Personen des öffentlichen Rechts", *Archiv des öffentlichen Rechts* 125 (2000), pp. 551-586, as well as Martin Nettesheim, "Kompetenzen", in: Armin von Bogdandy (ed.), *Europäisches Verfassungsrecht*, 1st ed. 2003, pp.

endowed with supporting management tasks[64]. Their independence is mainly a budgetary one to avoid uncertainties in responsibility within the Commission's budget[65].

The reason for the creation of an independent institution for the support of research is obvious. Starting from an idea of freedom of research, which alone can lead to convincing research results, there is an urgent need for structures that are free from (not only undue) political influence[66]. The inaugurating decision of the ERC[67] therefore provides (in its Article 4 (5)) that the members of its steering Scientific Council

> "... shall carry out their tasks independently of any outside influence."

Although "The Scientific Council shall operate in an autonomous and independent manner, Article 5 (1)", there is some tension with respect to its relationship with the Commission, as (according to Art. 5 (4)) it

> "... shall be accountable to the Commission, maintain continuous close liaison with it and the dedicated implementation structure, and establish any necessary arrangements for this."

415-477 at pp. 463 *et seqq*. In detail Matthias Ruffert, "Verselbständigte Verwaltungseinheiten: Ein europäischer Megatrend im Vergleich", in: Hans-Heinrich Trute/Thomas Groß/Hans Christian Röhl/Christoph Möllers (eds.), *Allgemeines Verwaltungsrecht – zur Tragfähigkeit eines Konzepts*, 2008, pp. 431 *et seqq.* at pp. 440 *et seqq.* and 446 *et seqq.*

64 Legal basis: Council Regulation (EC) No 58/2003 of 19 December 2002 laying down the statute for executive agencies to be entrusted with certain tasks in the management of Community programmes, [2003] O.J. L 11/1. Further Art. 54 and 55 of the Council Regulation (EC, Euratom) No 1605/2002 of 25 June 2002 on the Financial Regulation applicable to the general budget of the European Communities, [2002] O.J. L 248/1. On these provisions see Paul Craig, "The Constitutionalization of Community Administration", *Jean Monnet Working Paper* 3/2003.

65 Cf. Paul Craig (supra note 63), p. 37, with recourse to the Independent Experts' Second Report. Further Helmuth Schulze-Fielitz (supra note 63), at p. 106: factually subordinated authorities of the Commission.

66 On the creation of that Council cf. Armin von Bogdandy and Dietrich Westphal, "Der rechtliche Rahmen eines autonomen Europäischen Wissenschaftsrates", *Wissenschaftsrecht* 37 (2004), pp. 224-255. Cf. further Josef Franz Lindner (supra note 42), pp. 63 *et seqq.*

67 Commission Decision 2007/134/EC of 2 February 2007, [2007] O.J. L 57/14, establishing the European Research Council.

The operative business of the ERC is still limited to the emission of grants to well-established leading researchers (advanced grants) as well as to younger, promising researchers (starting grants). It should however be noted that the grants are of considerable pecuniary weight and that the research projects supported are not mainly oriented towards economic application but are undertaken in areas of edge-cutting, basic research. It should also be noted that the ERC is not considered to be a regulatory agency of the EU, but that its main purpose is the establishment of an institutional structure within the programme "ideas" that is part of the Seventh Framework Programme[68]. For administrative purposes, it is accompanied by the Executive Agency of the European Research Council whose main task is the budgetary supervision of all expenditure within the programme "ideas"[69].

(2) European Institute of Innovation and Technology (EIT)

Contrary to the ERC, the EIT is completely focused upon research for economic growth. It is apparently designed to follow the success of the famous Massachusetts Institute of Technology (MIT). Article 2 of its inaugurating regulation 294/2008[70] provides:

> "The EIT's objective is to contribute to sustainable European economic growth and competitiveness by reinforcing the innovation capacity of the Member States and the Community. It shall do this

---

[68] Decision of the European Parliament and of the Council (EC) No 1982/2006 of 18 December 2006, concerning the Seventh Framework Programme of the European Community for research, technological development and demonstration activities (2007-2013), [2006] O.J. L 412/1 (pp. 28 *et seq.*); Matthias Ruffert (supra note 40), Art. 187 para. 8; Henning Eikenberg (supra note 40), Art. 171 para. 40.

[69] Legal basis is Council Regulation (EC) No 58/2003, [2003] O.J. L 11/1, laying down the statute for executive agencies (19 December 2002) (cf. Matthias Ruffert (supra note 40), Art. 298, para. 9): Commission Decision (2008/37/EC) of 14 December 2007, setting up the 'European Research Council Executive Agency' for the management of the specific Community programme 'Ideas' in the field of frontier research in application of Council Regulation (EC) No 58/2003, [2008] O.J. L 9/15.

[70] Regulation of the European Parliament and of the Council (EC) No 294/2008 of 11 March 2008, establishing the European Institute of Innovation and Technology, [2008] O.J. L 97/1.

by promoting and integrating higher education, research and innovation of the highest standards."

That integration shall be achieved by creating so-called Knowledge Innovation Communities (KICs), clusters of universities, other research organisations and private (or public) undertakings that co-operate intensively with the support of the EIT[71]. In 2009, the three first KICs were selected which operate in the fields of climate change, sustainable energy and the information and communication society[72]. The EIT is designed like a regulatory agency (though not designated as such), and it is revealing that its legal basis is not Article 187 TFEU[73] or any other provision in the research chapter, but Article 157 (3) ECT (now Article 173 TFEU), the provision on the EU's industrial policy[74]. It has yet to be seen whether the design of research in Europe will be transformed as profoundly by the EIT as is sometimes feared[75].

---

71 Article 6 of Regulation No. 294/2008 (supra note 70).

72 Cf. the official information available at http://eit.europa.eu/press/news-archive/single-view/article/the-european-institute-of-innovation-and-technology-eit-launches-its-first-three-knowledge-and-inn.html.

73 On this basis, the Commission is entitled to erect European Research Infrastructure Consortia (ERIC) on the basis of the Council Regulation (EC) No 723/2009 of 25 June 2009 on the Community legal framework for a European Research Infrastructure Consortium (ERIC), [2009] O.J. L 206/1; see Wissenschaftsrat, *Empfehlungen zur deutschen Wissenschaftspolitik im Europäischen Forschungsraum*, 2010 (Drucksache 9866-10), at p. 64. These infrastructures are still in the planning phase, see European Commission, *Work Programme 2011, Capacities, Part 1, Research Infrastructures*, Doc. COM(2010)4903 of 19 July 2010, available at http://www.eubuero.de/arbeitsbereiche/infrastrukturen/Download/dat_/fil_5225.

74 On that policy see Matthias Ruffert, "Industriepolitik: Staatsdirigismus in der Marktwirtschaft?", in: Jürgen F. Baur/Otto Sandrock/Boris Scholtka/Amos Shapira (eds.), *Festschrift für Gunther Kühne zum 70. Geburtstag*, 2009, pp. 1021-1036.

75 Cf. the critical assesment by Josef Franz Lindner (supra note 42), p. 67 *et seq.*

(3) European University Institute (EUI)

In a separate Convention of 1972 (renewed 1992)[76], the Member States of the then European Community have created the European University Institute in Florence as an independent universitary institution for research and academic teaching. In this context, the College of Europe in Bruges, which is involved in academic teaching only has to be mentioned. Both are not agencies of the EU, but independently created public international legal bodies.

dd) Institutional Structures of Governance outside the Treaties

The same holds for a series of institutions in which the EU is a central or the only actor. In this respect, the following have to be mentioned:

- *EIROforum*[77], a collaboration between eight European intergovernmental scientific research organisations that are responsible for infrastructures and laboratories:
- the European Organization for Nuclear Research (CERN)[78],
- the European Southern Observatory (ESO)[79],
- the European Molecular Biology Laboratory (EMBL)[80],

---

[76] Legal basis is an administrative agreement of the contracting states Belgium, Germany, France, Italy, Luxembourg and the Netherlands in 1972, [1976] O.J. C 29/1, revised in 1992, see http://www.eui.eu/Documents/AboutEUI/Convention/ConsolidatedConventionRevising.pdf.

[77] Cf. http://www.eiroforum.org/.

[78] Convention for the establishment of a European organization for nuclear research (1953), available at: http://cdsweb.cern.ch/record/330625; Commission Decision 2009/488/EC, Euratom of 11 June 2009 on the conclusion of a Memorandum of Understanding between the European Commission and the European Organization for Nuclear Research (CERN), [2009] O.J. L 161/13 (replaces the Administrative Arrangement of 10 October 1994). Cf. for more information Felicitas Pauss, "Das CERN: Beispiel einer erfolgreichen europäischen Forschungsinstitution", in: Gerhart von Graevenitz and Jürgen Mittelstraß (eds.), *Das Projekt Europa: Rechts-, forschungs- und kulturpolitische Aspekte*, 2009, pp. 89-98; www.cern.ch.

[79] ESO was established by an intergovernmental Convention signed at Paris on 5 October 1962 by the governments of the Federal Republic of Germany, the Kingdom of Belgium, the French Republic, the Kingdom of the Netherlands and the Kingdom of Sweden, Cf. the ESO Convention BGBl. II 1965, pp. 43-67. Cf. for more information www.eso.org.

- the Institut Laue-Langevin (ILL)[81],
- the European Synchrotron Radiation Facility (ESRF)[82],
- the European Space Agency (ESA)[83],
- the European Fusion Development Agreement and the Joint European Torus (EFDA-JET)[84],
- the European X-ray Laser Facility (XFEL)[85]
- further the International Thermonuclear Experimental Reactor (ITER)[86],

---

80 Agreement about the establishment of EMBL: 954 U.N.T.S. 351 (1973), available at http://www.embl.de/aboutus/general_information/organisation/hostsite_agreement/. Cf. for more information www.embl.org.

81 The ILL was founded on 19 January 1967 with the signing of an agreement between the governments of the French Republic and the Federal Republic of Germany (cf. BGBl. II 1967, pp. 2430-2433). Cf. for more information www.ill.eu.

82 Convention Concerning the Construction and Operation of a European Synchrotron Radiation Facility (1988) and Annexes I-IV to the Convention (I: Statutes of the ESRF (1988); II: Target Specifications for Phase I; III: Estimated Annual Incidence of Expenditure; IV: Site Plan), available at: http://www.esrf.eu/AboutUs/Documentation/CompanyInfo/KeyDates/History/Convention; cf. for further information: ESFR, *A Light for Science*, available at http://www.esrf.eu/files/Brochures/ESRF-brochure-en.pdf.

83 http://www.esa.int/esaCP/index.html. Cf. for more information, e.g., the Annual Report 2008, available at http://esamultimedia.esa.int/multimedia/publications/Annual Report 2008/pageflip.html.

84 http://www.jet.efda.org/. Cf. European Fusion Energy Research: European Commission, *Euratom R&D: science & technology for energy – A true European Research Area*, 2007, available at http://www.jet.efda.org/wp-content/uploads/EC-Euratom-RD-2007.pdf.

85 Cf. http://www.xfel.eu/en/; Convention concerning the Construction and Operation of the European X-ray Laser Facility, available at http://www.xfel.eu/sites/site_xfel-gmbh/content/e63617/e79992/e68645/convention_english_eng.pdf; and Final Act, available at http://www.xfel.eu/sites/site_xfel-gmbh/content/e63617/e79992/e68656/final_act_english_eng.pdf, both signed on 30 November 2009.

86 Cf. the Agreement on the Establishment of the ITER International Fusion Energy Organization for the Joint Implementation of the ITER Project, [2006] O.J. L 358/62; Council Decision 2007/198/Euratom of 27 March 2007 establishing the European Joint Undertaking for ITER and the Development of Fusion Energy and conferring advantages upon it, [2007] O.J. L 90/58. Cf. also

- the European Research Coordination Agency (EUREKA), a pan-European research and development funding and coordination organization[87],
- the legal framework for European Research Infrastructure Consortia (ERIC)[88].

IV. Networks

1. Networks of Universities

a) University Co-Operation

It has almost become a platitude, that the somehow clear structure of States and International Organisations at global level has to be complemented by networks of sub-state entities as well as private actors[89]. In the field of science, such sub-state entities are to be found in universities, whatever their legal status (public or private) in the different jurisdictions[90].

---

the Communication from the Commission to the European Parliament and the Council, *ITER status and possible way forward*, COM(2010)226 final. Cf. for more information www.iter.org.

87 Hannover Declaration – Declaration of Principles relating to EUREKA, available at http://www.eurekanetwork.org/c/document_library/get_file?uuid=1b92be16-ec94-4a7e-a8d1-6dd40e4fb318&groupId=10137; Resolution on EUREKA, [1988] O.J. C 167/462; Resolution on the EUREKA project and the European Technological, [1985] O.J. C 352/98.

88 Council Regulation (EC) No 723/2009 of 25 June 2009 on the Community legal framework for a European Research Infrastructure Consortium (ERIC), [2009] O.J. L 206/1. Cf. also Wissenschaftsrat, *Empfehlungen zur deutschen Wissenschaftspolitik im Europäischen Forschungsraum*, 2010 (Drucksache 9866-10), at p. 64.

89 This development was enhanced if not triggered by the seminal work of Anne-Marie Slaughter, *A New World Order*, 2004. See also in this context Kal Raustiala, "The Architecture of International Cooperation: Transgovernmental Networks and the Future of International Law", *Virginia Journal of International Law* 43 (2001), pp. 1-92.

90 Cf. on the importance of such network-structures Wissenschaftsrat (supra note 88), at pp. 98 *et seq*.

The idea of academic research and teaching in universities has many historical sources, amongst which the activities of the brothers *Humboldt*, founding fathers of Berlin University in 1810, is certainly the most prominent one. Today, universities are a universally spread form of research institutions. Their legal status within and amongst the different jurisdictions varies considerably. They may be corporations under private law (with a strong business orientation), public bodies or even subentities of the State. If international co-operation between universities is scrutinised, such divergence in their legal status has to be considered.

As a matter of fact, the co-operation in research is indispensable for modern universities. The 274 German universities alone have established, according to their association (*Hochschulrektorenkonferenz*) about 19.000 transboundary co-operations with about 4.000 foreign universities. For this study, two examples have been scrutinised empirically[91]. Thus, the Friedrich-Schiller-University Jena has co-operative relationships with 75 foreign partners, and the University of Heidelberg about 300 universities[92]. Many of the co-operation agreements (but not all of them) comprise rules on the exchange of students and academic staff, others are limited to express the intention of close co-operation in scientific matters. The designations of the co-operation agreements vary widely, from memorandum, accord and agreement to contract, declaration or letter of understanding. It is highly doubtful that the said agreements are to be considered to be legally binding, and in some contracts the binding force is even expressly denied. As universities are not subjects of international law, we can by no means speak about international treaties or even apply the Vienna Convention on the Law of treaties.

However, the mere fact that the parties do not consider their cooperative relationships as legally binding and that the application of public international law strictly speaking is excluded does by no means trigger the irrelevance of such networks in our analysis. It is important to see that networks are being erected at a sub-national scale which operate regularly and on a reliable basis. Some of those networks play an important role in academic co-operation such as the Coimbra Group as the core network of traditional European universities[93]. Further inten-

---

91 The relevant documents are on file with the authors.

92 Cf. http://www.uni-heidelberg.de/international/uni-hd-international.html.

93 www.coimbra-group.eu; Bernd Wächter, *Handbook of European Associations in Higher Education: A Practical Guide to Academic Networks in Europe*

sive co-operations are created within the EU in the programme "Erasmus".[94]

b) The United Nations University

The United Nations have taken up the idea of university research and teaching and have established a university operated by the United Nations itself[95]. Its main campus is Tokyo, but it operates on a decentralised basis and is working with research institutions in all parts of the world. Its legal basis is a Charter contained in a Resolution of the General Assembly, and although such resolutions are considered to be legally binding only in very limited circumstances, there is no doubt about the legal soundness of the establishment of the United Nations University[96].

The activities of UNU are concentrated upon research. However, there is also an e-learning programme[97].

---

and Beyond, 2000, pp. 32-36; Jürgen Barkhoff and Helmut Eberhart (eds.), *Networking across Borders and Frontiers*, 2009, pp. *7 et seq*; "Der Europäische Wissensspeicher: Galerie der Genies – Ein Kunst- und Ausstellungsobjekt", *exposition catalogue on the occasion of the annual conference of the Coimbra Group in Jena*, 14-17 May 2008, p. 4.

[94] "Council Decision of 14 December 1989 amending Decision 87/327/EEC adopting the European Community action scheme for the mobility of university students (Erasmus) (89/663/EEC)", O.J. [1989] L 395/23; "Legislative Resolution embodying the opinion of the European Parliament on the proposal from the Commission to the Council for a decision relating to the adoption of a Community action programme in the field of vocational training and technological change (EUROTECNET II)", O.J. [1989] 323/172. "European Parliament legislative resolution of 21 October 2008 on the proposal for a decision of the European Parliament and of the Council establishing an action programme for the enhancement of quality in higher education and the promotion of intercultural understanding through cooperation with third countries (Erasmus Mundus) (2009-2013)", O.J. [2010] C 15 E/122. Cf general information: http://ec.europa.eu/education/lifelong-learning-programme/doc80_en.htm.

[95] See http://unu.edu/about/.

[96] GA Res. 3081 (XXVIII) of 6 December 1973.

[97] Cf. http://onlinelearning.unu.edu/en/.

2. Networks of Research Institutions other than Universities

a) Institutions Involved

Universities are not the only research institutions at national level. In many countries, independent research institutes and their associations are performing a considerable part of research efforts. Of course, such institutions can create transnational relationships, and they in fact do so. Again, we are faced here with a network structure similar to the university networks and below interstate co-operation[98].

Throughout the countries of the world, the variety of commitment in research is considerable. Sometimes, national academies are of utmost importance, and they are not only supporting research undertaken elsewhere, but they also perform own research, as in the U.S. (National Academy of Sciences) or France (Académie des Sciences). In other countries, independent institutions have been created to support scientific research by means of budgetary funding, project support or the issuance of grants, such as the Royal Society in the United Kingdom or in Germany the German Research Foundation (Deutsche Forschungsgemeinschaft).

b) The International Council for Science (ICSU)

aa) Creation and Membership

As early as 1931, such research and research funding institutions founded the International Council of Scientific Unions (ICSU) as a successor to the International Association of Academies (IAA; 1899-1914) and the International Research Council (IRC; 1919-1931), renamed International Council for Science in 1998 but keeping the acronym ICSU. As a non-profit, non-governmental organisation[99] ICSU is the most important knot in the international network of research organisations.

Today, ICSU comprises 121 national science organisations designated as National Scientific Members. As far as the organisation at national level is concerned, ICSU is open to any institutional design:

---

98 Cf. above 1.

99 Nr. 1 and 2 of ICSU Statutes and Rules of Procedure (2005), available at http://www.icsu.org/Gestion/img/ICSU_DOC_DOWNLOAD/216_DD_FILE_Statutes_October_2005.pdf.

> "8. A National Scientific Member shall be a scientific academy, research council, scientific institution or association of such institutions. Institutions effectively representing the range of scientific activities in a definite territory may be accepted as National Scientific Members, provided they can be listed under a name that will avoid any misunderstanding about the territory represented, and have been in existence in some form for at least 4 years.
>
> 9. The scientists of more than one nation may form a scientific body (academy, research council, etc.) for application as a National Scientific Member. No organization of scientists may adhere through more than one national membership."[100]

Further, 30 international scientific unions are part of ICSU.

> "7. A Scientific Union Member shall be an international non-governmental organization devoted to the promotion of activities in a particular area of science and shall have been in existence for at least 6 years."[101]

Such non-governmental organisations have been created for the promotion of science in a certain field mainly through the exchange of scholars, scientists and their ideas for a long time, and they merit separate scrutiny beyond their mere involvement in ICSU (below 3.). Within ICSU, they are joined by a considerable number of scientific associates[102].

bb) Activities

ICSU is supporting international research activities and interdisciplinary initiatives for research. A main field of support is geosciences including research on oceans, climate change, environmental problems and the polar regions. In formal terms, its main activities can be categorised as follows:

- *Strategic planning:* ICSU undertakes a strategic planning process to shape its global approach towards research policy and activity[103].

---

[100] ICSU Statutes and Rules of Procedure (2005), (supra note 99).

[101] ICSU Statutes and Rules of Procedure (2005), (supra note 99).

[102] Nrs. 11-15 ICSU Statutes and Rules of Procedure (2005), (supra note 99).

[103] Cf. its strategic plans for 2006-2011 and 2012-2017 (briefing document) available at http://www.icsu.org/Gestion/img/ICSU_DOC_DOWNLOAD/863_DD_FILE_ICSU_Strategic_Plan.pdf (2006-2011) and http://www.icsu.org/

- *Capacity building:* ICSU undertakes considerable efforts to establish a good scientific infrastructure comprising optimum facilities and working conditions. To this purpose, an ICSU grants programme has been established. International scientific exchange is promoted by a Visiting Scientists programme (cooperating with the Third World Academy of Sciences [TWAS], UNESCO and the Institute of Advanced Studies of United Nations University [UNU/IAS]). Specific information had originally been provided by the ICSU Committee on Capacity Building in Science (CCBS, 1993-2006), whose work is now continued under the aegis of the Inter-Academy Panel (IAP)[104]. An important ICSU-Activity is Science Education[105].
- *Data and Information:* A central aim of ICSU is the access to data and information from different disciplines, across national boundaries. Topics in this respect are new challenges relating to commercial value, accessibility, and security of scientific data. An ad hoc Strategic Coordination Committee on Information and Data (SCCID) has been created for the sound performance of that task, and ICSU actively participated in the 2001 World Summit on the Information Society. Intellectual property issues are extensively dealt with in this respect. The number and variety of sub-bodies and co-operative structures in the field of data and information is impressive[106].

Gestion/img/ICSU_DOC_DOWNLOAD/2242_DD_FILE_Briefing_paper_2nd_Strategic_Plan.pdf (2012-2017).

104 Cf. ICSU Capacity Building in Science http://www.icsu.org/1_icsuinscience/CAPA.html; IAP Capacity Building for Young Academies http://www.interacademies.net/CMS/Programmes/3128.aspx.

105 Cf. ICSU Annual Report 2009, p. 6.

106 Current ICSU activities in relation to Data and Information:

- the International Network for the Availability of Scientific Publications (INASP), an interdisciplinary body, which was established to help bridge the information divide between developed and developing countries;
- the International Council for Scientific and Technical Information (ICSTI), an important associate member of ICSU;
- several interdisciplinary bodies, whose principal focus is the management and use of large scientific data sets: Committee on Data for Science and Technology (CODATA), Panel of the World Data Centres (WDC), Federation of Astronomical and Geophysical Data Analysis Services (FAGS);
- the Scientific Committee on Frequency Allocations for Radio Astronomy and Space Science (IUCAF), that has a very focussed co-ordination role

- *Environment and sustainable development:* Since the Johannesburg World Summit 2002 at the latest, ICSU has begun to undertake own research activities in the environmental field[107].
- *Grants:* Finally, ICSU is issuing grants for the support of scientific activities world-wide[108].

As it is often the case in international governance, the core budget of the organisation is relatively small – € 3.5 million –, but the overall expenditure on ICSU programmes cannot easily be detected[109]. Further, it is important to note that ICSU is strongly committed to the idea of the universality of science which is described as follows:

> "This principle embodies freedom of movement, association, expression and communication for scientists, as well as equitable access to data, information and research materials. In pursuing its objectives in respect of the rights and responsibilities of scientists, the International Council for Science (ICSU) actively upholds this principle, and, in so doing, opposes any discrimination on the basis of such factors as ethnic origin, religion, citizenship, language, political stance, gender, sex or age. ICSU shall not accept disruption of its

---

relating to the activities of the WDC and FAGS, and the Global Observing Systems (GTOS, GOOS, GCOS), which are co-sponsored by ICSU;

- either individually or working together, many members of ICSU are very active with regards to scientific data, e.g. in 2002 an ad hoc Inter-Union Bioinformatics Group produced an authoritative report and recommendations regarding biological databases (Cf. ://www.icsu.org/1_icsuinscienchttpe/DATA_Other_1.html).

To provide broad expertise and advice to ICSU in the field of Scientific Data and Information the ICSU Strategic Coordinating Committee for Information and Data (SCCID) has been established. This committee acts as an interface between scientists and data and information professionals that can advise on the data needs and possible solutions for existing and new ICSU programmes and other international initiatives (Cf. http://www.icsu.org/5_abouticsu/STRUCT_Comm_Adhoc_SCCID.html).

[107] Cf. the extensive ISCU environment portfolio http://www.icsu.org/1_icsuinscience/ENVI_Portfolio_1.html.

[108] http://www.icsu.org/1_icsuinscience/GRANTS_1.html; cf. the Review of the ICSU Grants Programme 2001-2006 – Report of a CSPR Review Committee http://www.icsu.org/Gestion/img/ICSU_DOC_DOWNLOAD/1390_DD_FILE_Grants_Review_02-2007.pdf.

[109] Cf. the core funding http://www.icsu.org/5_abouticsu/FINAN_CoreFund_1.html#IE.

own activities by statements or actions that intentionally or otherwise prevent the application of this principle.

cc) Structure

Although being shaped as an NGO, the structure of ICSU follows the typical design of an International Organisation[110]. Its plenary organ is the General Assembly (GA) responsible for the main decisions, whereas its Executive Board (EB), composed of six Officers (President and other leaders) and eight Ordinary Members is in charge of implementing ICSU's policies. The work of these bodies is supported by various Committees, Interdisciplinary Bodies and Joint Initiatives on the one hand, the Secretariat (based in Paris) and four Regional Offices on the other hand.

dd) ICSU, InterAcademy Panel on International Issues (IAP) International Academy Council (IAC)

It should be noted that ICSU is not the only organisation managing international co-operation of national academies of science. This task is promoted further by the InterAcademy Panel on International Issues (IAP), founded in 1993, which in 2000 created the International Academy Council (IAC) as a foundation under Dutch law. IAP and IAC both issue papers on research policy issues of general concern for mankind.

ee) InterAcademy Medical Panel (IAMP)

The InterAcademy Medical Panel (IAMP) comprises academies of science in the medical field.

---

[110] Cf. Christian Walter and Matthias Ruffert (supra note 12), para. 288. Cf. Nrs. 16 *et seq.* ICSU Statutes and Rules of Procedure (2005), (supra note 99).

3. Professional Bodies

a) International Council of Academies of Engineering and Technological Sciences (CAETS)

In the context of international governance of science, professional bodies bringing together researchers within particular scientific communities must be mentioned. Their aim is to organise co-operation in scientific matters following the general research focus of a professional community. A prominent example of such a community is the International Council of Academies of Engineering and Technological Sciences (CAETS)[111]. Beyond mere research co-operation, such international non-governmental organisations may influence the standard of research and the interaction between researchers and politics or society by statements, guidelines and participation in conferences or other forms of dialogue. This is clearly stated in the CAETS byelaws[112]:

> *Article 1 – Objectives*
>
> Consistent with its Articles of Incorporation and in support of its mission, CAETS will:
>
> (a) Provide an independent nonpolitical and non-governmental international organization of engineering and technological sciences academies, prepared to advise governments and international organizations on technical and policy issues related to its areas of expertise;
>
> (b) Contribute to the strengthening of engineering and technological activities in order to promote sustainable economic growth and social welfare throughout the world;
>
> (c) Foster a balanced understanding of the applications of engineering and technology by the public;
>
> (d) Provide an international forum for discussion and communication of engineering and technological issues of common concern;
>
> (e) Foster cooperative international engineering and technological efforts through meaningful contacts for development of programs of bilateral and multilateral interest;
>
> (f) Encourage improvement of engineering education and practice internationally;

---

[111] For a comprehensive assessment: *CAETS, The First Thirty Years 1978-2008*, 2008.

[112] Available at: http://www.caets.org/cms/15/7139.aspx.

(g) Foster establishment of additional engineering academies in countries where none exist; and

(h) Undertake other projects, programs, and activities not inconsistent with section 501 (c) (3) of the Internal Revenue Code and any applicable law of the District of Columbia.

b) World Medical Association (WMA[113])

In the field of medical research and practice, the declarations of the World Medical Association (WMA) are of utmost importance. The WMA Declaration of Helsinki – Ethical Principles for Medical Research Involving Human Subjects (1964, last amended 2008) is certainly the most important international document concerning research in human medicine[114]. Similar to CAETS, the WMA is an international non-governmental organisation composed of (currently about 80) national medical doctors' associations. Again, the institutional structure resembles the traditional structures of international organisations: General Assembly, WMA Council headed by a President and Secretariat[115].

V. Other Non-State Actors

Pugwash is a village in the North East of Canada with barely 800 inhabitants[116]. It has become famous, though, hosting the Pugwash Con-

---

113 Cf. the constitution of the World Medical Association: http://www.worldmedicalassociation.org/05_Constitution/0101_Impressum.php.

114 Cf. Erwin Deutsch and Jochen Taupitz, "Einführung: Forschungsfreiheit und Forschungskontrolle in der Medizin – zur geplanten Revision der Deklaration von Helsinki", in: id. (eds.), *Freedom and Control of Biomedical Research – The planned Revision of the Declaration of Helsinki*, 2000, pp. 1-6 at p. 3; Jana Straßburger, "Die Inkorporation der Deklaration von Helsinki in das ärztliche Berufs- und Standesrecht – Verfassungsrechtliche Aspekte", *Medizinrecht* 2006, pp. 462-471 at p. 462.

115 Cf. the structure of the WMA: http://www.wma.net/en/60about/30structure/index.html.

116 Cf. the population of Pugwash in 2006 the Nova Scotia Community counts http://www.gov.ns.ca/finance/communitycounts/profiles/community/defaultp.asp?gnum=com1110&gnum2=com1110&gname=&gview=2&glevel=com&gtype=&ptype=&gsel=&acctype.

ferences on Science and World Affairs since 1957. These Conferences were initially directed mainly against the development and proliferation of weapons of mass destruction as described in the Russell-Einstein-Manifesto of 1955[117]:

> In the tragic situation which confronts humanity, we feel that scientists should assemble in conference to appraise the perils that have arisen as a result of the development of weapons of mass destruction, and to discuss a resolution in the spirit of the appended draft...

In this context, their representative *Józef Rotblat* was awarded the Nobel Peace Prize in 1995. The Conferences have crystallised into an international non-governmental organisation dealing with ethical matters of natural scientists, motivated by its founders who had formerly been involved in the creation of nuclear weapons[118].

Other NGOs might touch upon scientific activities in the context of their own thematic range which is not scientific in the first place. For example, the World Wildlife Fund for Nature (WWF) issues declarations against research activities that may be harmful to certain animal species, e.g. against "scientific whaling"[119].

VI. Results

Summing up the most salient features of the institutional international governance of science, a systematic or even harmonic picture cannot easily be drawn. At any rate, such institutional variety does by no means come as a surprise at global level, given that the organisation of science is manifold at domestic level – for the very reason that the free development of scientific efforts is the centre of interest. What appears to be even chaotic at first sight is due to the – rightly – limited influence of government in the field, so that governance is more complicated to be designed.

---

117 Available at: http://www.pugwash.org/about/manifesto.htm.

118 Cf. the Pugwash Mission Statement for the Eleventh Quinquennium: 2007-2012, available at: http://www.pugwash.org/about/mission.htm.

119 WWF respective to th misusage of science for commercial benefits: *Japanese Scientific Whaling: Irresponsible Science, Irresponsible Whaling*, available at: http://wwf.panda.org/about_our_earth/all_publications/?13793/Japanese-Scientific-Whaling-Irresponsible-Science-Irresponsible-Whaling.

A most eminent observation is the interrelationship and entanglement of different categories of institutions. States, inter- and supranational organisations are co-operating with each other and with non-governmental entities. Institutions designed by legal means strictly speaking are related to legally informal associations created on a voluntary basis. It may even be doubted whether organisations at that level are governed by legal provisions even in the broadest sense. This, however, can only be discerned when creation and implementation of principles and rules is taken into account.

E. Governance Mechanisms

I. Rulemaking, Implementation and Management

From a public international legal point of view, the creation of international rules is one of the most controversial – and also interesting – problems discussed at theoretical level in the area of global governance. The obvious limits of Article 38 (1) of the ICJ-Statute[1], the issue of the binding force of rules and (related to this) the bifurcation between legal and social norms are just keywords to denote that theoretical controversy which is far from having been resolved[2]. Reconsidering the plurality of institutional actors described under D., it may be submitted that there is at least a minimum categorisation of clear and contentious cases in international legal theory: networks of sub-state entities or international non-governmental organisations are subject to scrutiny by public international lawyers, and although their status in public international law is not clear in detail, their existence is at least undoubted as distinct categories. Contrary to this, uncertainty with respect to rules and lawmaking could not be greater, hidden behind smoke screens denominated such as "soft law" or "self-regulation".

Moreover, in the framework of global administrative law, the analysis of the mere formation of rules is insufficient and must be completed by identifying distinctive tools of implementation and management. This also applies from the perspective of global governance which focuses upon regulatory structures (including implementation and management) rather than single institutions or rules. It is theoretically most unclear, though, how rulemaking procedures and their results are re-

1 Which had been made out even when the Article was drafted, as shown by Alain Pellet, in: Andreas Zimmermann/Christian Tomuschat/Karin Oellers-Frahm (eds.), *The Statute of the International Court of Justice,* 2006, Article 38, para. 82.

2 On these crucial issues see Matthias Goldmann, "Inside Relative Normativity: From Sources to Standard Instruments for the Exercise of International Public Authority", in: Armin von Bogdandy/Rüdiger Wolfrum/Jochen von Bernstorff/Philipp Dann/Matthias Goldmann (eds.), *The Exercise of Public Authority by International Institutions,* 2010, pp. 661-711; José E. Alvarez, *International Organisations as Law-makers,* 2005, at pp. 588 *et seq.*

lated to mechanisms of procedural implementation or effective management. Rules may create viable regimes *per se*, management schemes may operate without or instead of rules, and such displacements or entanglements may create legal problems of their own. Even without considering a global doctrine of separation of powers to be necessary – a point that shall be left aside here – it is obvious that rulemaking, implementation and management and their interrelationship must fulfil core legal requirements which are, in the field of science, formulated according to the orientation towards free science (supra C.) above all.

As a matter of course, the present analysis does not ignore these theoretical points. On the contrary, it is along these lines that their solution as problems to global administrative law is thought in analysing the international governance of science to achieve further scholarly advancement in this field. To focus upon its actual governance mechanisms, the theoretical questions are not discussed in the first place, but the study starts here with a structured empirical view on the international rules in existence for the governance of science together with their implementation and management schemes. This view is the emanation of a long-term collection of such rules undertaken within our interdisciplinary research project[3]. To conserve sight within the dense collection of texts, it is useful to start with a first categorisation related neither primarily to the legal quality or designation of the normative texts as collected nor to their legal effects in these texts but to the subject matters in which they have been issued as related to science. In this respect, three interrelated categories can be detected, which demonstrate the purposes and necessities of rulemaking, implementation and management within the broad field of science. The interrelationship between the first and second is strongest; both are about ethical rules, the first with respect to the soundness of performing scientific activity as such (remember the *Hwang Woo-Suk*-case after the detection of falsification of data and results), the second with respect to other rights and values affected by scientific activities (remember the same case when *Hwang Woo-Suk* turned out having applied illegal methods receiving the object of his research). The third category may be linked to these ethical points but is concerned mainly with other issues around the support for research and the legal position of researchers.

---

[3] Cf. the data collection http://www.rewi.uni-jena.de/Fakult_auml_t/Professoren+_amp_+Dozenten/Universit_auml_tsprofessoren/Prof_+Dr_+Matthias+Ruffert/Forschung/Forschungsprofil/Globalisierung+und+Global+Governance/Elemente+eines+transnationalen+Wissenschaftsrechts/Database.html.

II. Governance Purposes in the Global Administrative Law of Science

1. Ethical Standards for Research and Their Implementation I: Sound Scientific Practice

To begin with an almost classical field of governance in science, there is consensus within the rules considered about a set of ethical rules applicable to all scientific activity to ensure the soundness of scientific practice[4]. Freedom always needs rules for the prevention of its abuse. Eventually, freedom of science would be impaired if it could conceal malpractice as the falsification of data or plagiarism. Therefore, all instruments referring to the activity of scientists contain rules relating to (1) the sound collection of data, (2) the correct performance of the scientific activity and (3) the accuracy of publication. In phase (1), the provenience of data and the methods of its detection must be made transparent and conserved for a certain time[5]. Phase (2) has to be undertaken in a transparent way as well[6], and in phase (3), the elaboration of one's own results shall go along with correct quotations of other results and opinions[7]. Any trace of plagiarism must be avoided by naming any con-

---

[4] Cf. Sebastian Steinecke, *Zur internationalen Governance der Wissenschaft*, 2010, manuscript pp. 357 *et seq.*

[5] Cf. Association of Social Science Researchers: Code of Ethics (1996); Code of Ethics for Researchers of the Academy of Sciences of the Czech Republic (2006); American Physical Society: Ethics & Values (1991); American Statistical Association: Ethical Guidelines for Statistical Practice (1999) etc. – examples are taken from the data collection.

[6] Cf. Statements on Ethics – Principles of Professional Responsibility, adopted by the Council of the American Anthropological Association (1971); Code of Ethics of the American Institute for Conservation of Historic and Artistic Works; Elements of a Code of Conduct for Ocean Fertilization Projects (2007); Ethical Guidelines for International Comparative Social Science Research in the framework of MOST; International Code of Conduct for Plant Germplasm Collecting and Transfer etc.

[7] Cf. Code of Ethics for Researchers of the Academy of Sciences of the Czech Republic (2006); Code of Ethics for Researchers of the Academy of Sciences of the Czech Republic (2006) etc.

tributor to a scientific publication. Sometimes, there is a duty to publish[8].

A more controversial field is the avoidance of a conflict of interests. Researchers are not only performing science in universities or other independent entities, but also in national or multinational companies, whose interest may influence the process and results of scientific research[9]. The pertinent rules are not consistent here. Sometimes special procedures for dealing with such conflicts of interest are established[10], sometimes the disclosure of conflicts of interest is required[11], sometimes particular review boards shall handle the conflict[12], but there are also rules requiring the withdrawal of the scientist(s) concerned from the relevant project or even its abandonment[13].

In the context of such rules, it is essential to mention the procedural safeguards established for the maintenance of the ethical standards as described and thus the pertinent methods of implementation. Formal procedures leading to a form of punishment of a scientist who contra-

---

8 Cf. Code of Ethics for Researchers of the Academy of Sciences of the Czech Republic (2006); Code of Ethics of the American Anthropological Association (1998); Statements on Ethics – Principles of Professional Responsibility, adopted by the Council of the American Anthropological Association (1971); World Medical Association Declaration of Helsinki – Ethical Principles for Medical Research Involving Human Subjects (1964) etc.

9 The main challenges to the freedom of science are seen in this field by Dieter Grimm, *Wissenschaftsfreiheit vor neuen Grenzen?*, 2007.

10 Committee on Publication Ethics: A code of conduct for editors of biomedical journals, available at http://publicationethics.org/code-conduct.

11 Council for International Organizations of Medical Sciences: International Ethical Guidelines for Biomedical Research Involving Human Subjects, Commentary on Guideline 2 "Ethical review committees" (2002), http://www.cioms.ch/publications/guidelines/guidelines_nov_2002_blurb.htm, and Nrs. xvi-xviii of the Appendix to the Additional Protocol to the Convention of Human Rights and Biomedicine, concerning Biomedical Research (ETS Nr. 195, 2005).

12 American Society of Human Genetics: Statement on Gene Therapy (2000), http://www.ashg.org/pages/statement_ajhg67.shtml.

13 American Statistical Association: Ethical Guidelines for Statistical Practice (1989, superceded 1999), http://www.amstat.org/about/ethicalguidelines.cfm; American Physical Society Guidelines for Professional Conduct (1991, superceded 2002), http://www.aps.org/policy/statements/02_2.cfm.

venes one of the rules are rare and reserved to severe forms of breach[14]. In the end, ordinary penal law can claim its place if we are faced with instances of fraud or bribery. More frequent, however, is the withholding or withdrawal of (public) funding[15]. In this respect, it is essential that the institutions deciding upon such sanctions or (only) giving advice in the ethical questions explained are not mislead in exerting undue influence upon the scientific process[16]. Therefore, it is mostly non-

---

[14] In a very large extent Computer Society of India: Code of Ethics, 1993 (Art. 7 of the Code of Ethics Procedure for Action against a Member for any Breach of the Code of Ethics: *"The findings of the Honours Committee would depend on the merits of each case and their recommendation to the ExecCom may be a) Honourable acquittal. b) Removal from membership."*). Cf. also Alexander von Humboldt Foundation: Rules of good scientific practice, procedures, and penalties in the event of malpractice (2007). Other examples for procedural safeguards (taken from the data collection): *Committee on Publication Ethics* (COPE) – "Pursuing Misconduct" of the Code of Conduct for Editors of Biomedical Journals (2004); American Historical Association – Art. 4 of the Statement on Standards of Professional Conduct (2005); American Institute for Conservation of Historic and Artistic Works –Art. 13 of the Guidelines for Practice (1994); Ergonomics Society – "Disciplinary Procedure" of the Code of Conduct (2008).

[15] Cf. German Research Foundation (Deutsche Forschungsgemeinschaft, DFG): Verwendungsrichtlinien. Sachbeihilfen mit Leitfaden für Abschlussberichte und Regeln guter wissenschaftlicher Praxis – Formblatt 2.01, p. 21, available at http://www.dfg.de/download/pdf/foerderung/rechtliche_rahmen bedingungen/gute_wissenschaftliche_praxis/self_regulation_98.pdf; NIH/DFG Research Career Transition Awards Program – Guideline for proposals, p. 2: "the DFG may [...] exclude those found responsible from the right to apply for DFG funds for a period of one to eight years, depending on the severity of the scientific misconduct; revoke funding decisions (completely or partially revoke approvals, demand the return of authorised funds, and the repayment of funds spent)", available at http://www.dfg.de/download/pdf/foerderung/programme/nih_dfg/guideline_for_proposal.pdf; Article 3 of the Rules of good scientific practice, procedures, and penalties in the event of malpractice of the Alexander von Humboldt Foundation (2007), available at http://www.humboldt-founda tion.de/pls/web/docs/F16253/regelnwisspraxis.pdf. None of the codes contained in the data collection institutionalises direct financial sanctions.

[16] Hans-Heinrich Trute, *Die Forschung zwischen grundrechtlicher Freiheit und staatlicher Institutionalisierung. Das Wissenschaftsrecht als Recht kooperativer Verwaltungsvorgänge*, 1994, pp. 300 *et seqq.*

governmental institutions that are established and empowered to exert the necessary control in processes of self-regulation.[17]

2. Ethical Standards for Research II: Bioethics

Self-regulatory procedures are viable with respect to intrinsic problems of scientific research, although repercussions of potentially unsound scientific practice upon society in general cannot be denied[18]. There are, however, other ethical questions which are affecting rights, values and interests beyond the scientific community from the outset. A particularly difficult area is research in biosciences affecting human dignity, life, health and/or self-determination.

In such a controversial field, simple and globally consented rules are not to be expected. However, two legal prescriptions have found overall acceptance:

Firstly, the prohibition of the reproductive cloning of human beings is enshrined in a series of documents such as the Convention for the Protection of Human Rights and Dignity of the Human Being with regard to the Application of Biology and Medicine: Convention on Human Rights and Biomedicine (1997) in the framework of the Council of Europe[19] (together with the Additional Protocols[20]), the Universal Dec-

---

[17] E.g. American Historical Association, American Institute for Conservation of Historic and Artistic Works, Council for International Organizations of Medical Sciences (CIOMS), International Association for the Study of Pain, International Society for Environmental Epidemiologists.

[18] Sebastian Steinecke, (supra note 4), manuscript p. 377.

[19] "Article 18 – Research on embryos *in vitro* (1) Where the law allows research on embryos *in vitro*, it shall ensure adequate protection of the embryo. (2) The creation of human embryos for research purposes is prohibited" (available at http://conventions.coe.int/Treaty/en/Treaties/Html/164.htm). On that convention see Eibe Riedel, "Global Responsibilities and Bioethics: Reflections on the Council of Europe's Bioethics Convention", *Indiana Journal of Global Legal Studies* 5 (1997), pp. 179-190.

[20] Additional Protocol to the Convention for the Protection of Human Rights and Dignity of the Human Being with regard to the Application of Biology and Medicine, on the Prohibition of Cloning Human Beings (ETS No. 168, available at http://www.coe.int/t/dg3/healthbioethic/Activities/03_Cloning_en/168ProtocolCloningER_en.pdf). Additional Protocol to the Convention on Human Rights and Biomedicine, Concerning Biomedical Research (ETS

laration on the Human Genome and Human Rights (11 November 1997) in the framework of UNESCO[21], the resolution of the World Health Organisation on Cloning in Human Reproduction[22], and can be considered as customary international law, although the UN efforts to establish an International Convention against the Reproductive Cloning of Human Beings have only lead to a General Assembly resolution to this effect so far[23].

Secondly, in all bioscientific research activity, the principle of informed consent has to be respected, i.e. human beings who are subject to research in medical, biological or other matters have to express their will

---

No. 195), available at http://conventions.coe.int/treaty/en/treaties/html/195.htm.

21 Article 11 "Practices which are contrary to human dignity, such as reproductive cloning of human beings, shall not be permitted. States and competent international organizations are invited to co-operate in identifying such practices and in taking, at national or international level, the measures necessary to ensure that the principles set out in this Declaration are respected."

22 Resolution 50.37 (14.5.1997): *Cloning in Human Reproduction*. Cf. the reports on implementation of resolution WHA50.37 concerning ethical, scientific and social implications of cloning in human health, available at http://apps.who.int/gb/archive/pdf_files/WHA51/ea6a1.pdf.

23 United Nations Declaration on Human Cloning, UN Doc. A/RES/59/280 of 8 March 2005; on this cf. Report of the Working Group established pursuant to General Assembly decision 59/547 to finalize the text of a United Nations declaration on human cloning, UN Doc. A/C.6/59/L. 27/Rev. 1; and the Report of the Sixth Committee on the International convention against the reproductive cloning of human beings, UN Doc. A/59/516 of 19 November 2004 with UN Doc. A/59/516/Add.1 of 24 February 2005.

Report of the Working Group established pursuant to General Assembly decision 59/547 to finalize the text of a United Nations declaration on human cloning, UN Doc. A/C.6/59/L.27 and Rev.1 (16 and 23 February 2005), available at http://daccess-dds-ny.un.org/doc/UNDOC/LTD/N05/240/64/PDF/N0524064.pdf?OpenElement and http://daccess-dds-ny.un.org/doc/UNDOC/LTD/N05/247/70/PDF/N0524770.pdf?OpenElement.

Cf. Mahnoush H. Arsanjani, "The Negotiations on a Treaty on Cloning: Some Reflections", in: Silja Vöneky and Rüdiger Wolfrum (eds.), *Human Dignity and Human Cloning,* 2004, pp. 145-165 and generally the article by Hans Lilie, "International Legal Limits to Human Cloning", in the same volume at pp. 125-132.

to participate on a basis of rational comprehension of the experiment or other research activity undertaken[24].

---

[24] *World Medical Association Declaration of Helsinki* (1964), available at http://www.wma.net/en/30publications/10policies/b3/17c.pdf;

Nr. 22: "Participation by competent individuals as subjects in medical research must be voluntary. Although it may be appropriate to consult family members or community leaders, no competent individual may be enrolled in a research study unless he or she freely agrees."

Nr. 24 "In medical research involving competent human subjects, each potential subject must be adequately informed of the aims, methods, sources of funding, any possible conflicts of interest, institutional affiliations of the researcher, the anticipated benefits and potential risks of the study and the discomfort it may entail, and any other relevant aspects of the study. The potential subject must be informed of the right to refuse to participate in the study or to withdraw consent to participate at any time without reprisal. Special attention should be given to the specific information needs of individual potential subjects as well as to the methods used to deliver the information. After ensuring that the potential subject has understood the information, the physician or another appropriately qualified individual must then seek the potential subject's freely-given informed consent, preferably in writing. If the consent cannot be expressed in writing, the non-written consent must be formally documented and witnessed."

Convention for the Protection of Human Rights and Dignity of the Human Being with regard to the Application of Biology and Medicine: Convention on Human Rights and Biomedicine (1997) of the Council of Europe:

Article 16 – Protection of persons undergoing research:

"Research on a person may only be undertaken if all the following conditions are met:

i. there is no alternative of comparable effectiveness to research on humans;
ii. the risks which may be incurred by that person are not disproportionate to the potential benefits of the research;
iii. the research project has been approved by the competent body after independent examination of its scientific merit, including assessment of the importance of the aim of the research, and multidisciplinary review of its ethical acceptability;
iv. the persons undergoing research have been informed of their rights and the safeguards prescribed by law for their protection;
v. the necessary consent as provided for under Article 5 has been given expressly, specifically and is documented. Such consent may be freely withdrawn at any time."

A series of other principles and rules along these basically uncontested dicta is also mainly beyond controversy, but not as firmly entrenched: the right to free choice of doctor as well as religious succor[25], agreement of the donors in the research on spare ova as a result of In-Vitro-Fertilisation[26], order of oocyte-reduction in reproductive medicine[27] as well as binding standards of animal protection in biomedical research [28]. The general public should be informed about plans to perform research activities[29]. The individual physician should bear the responsibility for his own research activity[30]. Trade in parts of the human body (including cells for reproduction or embryos) is prohibited[31].

Other rules fail to gain overall recognition so far, and they are even overtly contested due to different cultural attitudes[32]. Therapeutic cloning and manipulations of embryonic stem cells are prohibited in some jurisdictions only whereas there is a general permission of such research activity in other jurisdictions.

The procedural safeguards for the conservation of conflicting rights and values deserve particular attention, as methods of implementation have

---

25 World Medical Association: Declaration on the Rights of the Patient (2005).

26 World Medical Association: Statement on In-Vitro Fertilization and Embryo Transplantation (1987).

27 World Medical Association: Statement on Ethical Aspects of Embryonic Reduction (1995).

28 World Medical Association: Statement on Animal Use in Biomedical Research (2006).

29 Article 15 of the WMA Declaration of Helsinki – Ethical Principles for Medical Research Involving Human Subjects (2004), available at http://www.wma.net/en/30publications/10policies/b3/index.html.

30 Article 16 of the WMA Declaration of Helsinki – Ethical Principles for Medical Research Involving Human Subjects (2004).

31 Cf. World Medical Association: Resolution on the Non-Commercialisation of Human Reproductive Material (2003), available at http://www.wma.net/en/30publications/10policies/r1/index.html: prohibition of commercial transactions in human ova, sperm and embryos; World Medical Association: Statement concerning the Relationship between Physicians and Commercial Enterprises (2004), available at http://www.wma.net/en/30publications/10policies/r2/index.html: prohibition of the receiving of payments in cash or cash equivalents that are not directly linked or that are disproportionate.

32 Cf. the contributions in Part 1 of Silja Vöneky and Rüdiger Wolfrum (eds.) (supra note 23).

been established which are particular to the field and have gained overall acceptance. In national rules[33] as well as in diverse inter- and transnational instruments[34], ethics committees are established as independent

---

[33] Ethics Committees/Institutional Review Boards/Research Ethics Committees exist in almost all countries. Instead of all these see just the case studies that were chosen for the project: Europe (exemplary: Albania (Albanian Committee on Bioethics), Belgium (Comité consultatif de Bioéthique de Belgique), Denmark (Den Centrale Videnskabsetiske Komité, CVK), Estonia (Eesti Bioeetika Noukogu), Finland (Tutkimuseettinen Neuvottelukunta), France (Comité Consultatif National d'Ethique pour les sciences de la vie et de la santé), Georgia (National Bioethics Council of Georgia), Greece (ΕΘΝΙΚΗ ΕΠΙΤΡΟΝΗ ΒΙΟΗΘΙΚΗΣ), Great Britain (Nuffield Council on Bioethics), Ireland (Irish Council for Bioethics), Iceland (National Bioethics Committee of Iceland), Italy (Comitato Nazionale Italiano per la Bioetica, Consiglio Nazionale delle Ricerche), Lithuania (Lietuvos bioetikos komitetas), Luxembourg (Commission Consultative Nationale d'Ethique pour les Sciences de la Vie et de la Santé), Malta (Bioethics Consultative Committee), Netherlands (Gezondheidsraad), Norway (De nasjonale forskningsetiske komiteer), Austria (Bioethikkommission), Portugal (Conselho Nacional de Etica para as Ciencias da Vida), Russia (Russian National Committee on Bioethics), Sweden (Statens medicinsk-etiska råd), Switzerland (Commission nationale d'ethique pour la medicine humaine, Commission fédérale d'ethique dans la biotechnologie nonhumain), Slovakia (Central Ethic Committee of the Ministry of health), Slovenia (National Medical Ethics Committee), Czech Republic (Centralni Eticka Komise Pri Ministerstvu Zdravotnictvi Ceske Republiky), Hungary (Egészségügyi Tudományos Tanács Tudományos és Kutatásetikai Bizottsága), Cyprus (Cyprus National Bioethics Committee)), USA (President's Council on Bioethics), Japan (Bioethics Committee, Council for Science and Technology), China (Department of Research on Ethical, Legal and Social Implications at Chinese National Human Genome Center), Argentina (Argentina National Bioethics Advisory Commission), Brazil (Comissão Nacional de Éthica em Pesquisa), Nigeria (National Ethics Committee of Nigeria), Australia (Australian Health Ethics Committee), Saudi Arabia (National Committee for Medical and Bio-ethics), Egypt (Egyptian National Committee for Bioethics), India (Council of Medical Research), Indonesia (Health Research Committee), South Africa (Research Ethics Council and the Ethics Committee of the South African Medical Research Council).

[34] Cf. e.g. Art. 7 of the Madrid Declaration on Ethical Standards for Psychiatric Practice (1996), referring to "an appropriately constituted ethical committee"; the Committee on Ethics of the American Anthropological Association; Art. 4.1.7. of the Ethics Guidelines for Environmental Epidemiologists referring to Review Committees; the International Association for the Study of Pain Guidelines, referring to an "independent committee on human research" and an "appropriately constituted peer review committee"; the Panel of Eminent Ex-

bodies for the evaluation of research that might impede upon such rights or values[35]. Many of the rules provide that the committees shall consider the quality of research to be undertaken[36]. This is called into question by virtue of freedom of science, as the assessment of that quality should remain with the scientist. At any rate, the main emphasis of the committees' activity lies on the evaluation of ethical questions[37]. The committees may also monitor research and request monitoring information from the researcher[38].

It should finally be noted that biomedicine is not the only ethically relevant sphere of research. Scientific activity for military purposes also has an ethical aspect, but due to the allocation of the field to the sphere

---

perts on Ethics in Food and Agriculture of the *Food and Agriculture Organization* of the United Nations (FAO); the International Bioethics Committee (IBC), established by the UNESCO.

35 In the USA the first ethic commissions were established in the 1970s after several scandalous medical experiments and serious medical misconduct caused an enormous pressure to act: Since 1932 doctors carried out a long-term study on the course of untreated syphilis with 400 colored US-Americans. In the 1940s the discovery and the progressive introduction of penicillin into medicine would have allowed a medical treatment of the test persons with the prospect of cure. However the doctors involved did not start a medical treatment but continued their study until 1972 when it was finally stopped by the pressure of the informed public (see Gerd Brudermüller, "Ethikkommissionen und ethischer Diskurs", in: id. (ed.), *Angewandte Ethik und Medizin*, 1999, pp. 85-116).

In international law, Articles 9-12 of the Additional Protocol to the Convention on Human Rights and Biomedicine, concerning Biomedical Research (ETS Nr. 195, 2005), provide for the establishment of such committees.

36 For their legitimacy cf. Silja Vöneky, "Ethische Experten und moralischer Autoritarismus", in: id./Cornelia Hagedorn/Miriam Clados/Jelena von Achenbach (eds.), *Legitimation ethischer Entscheidungen im Recht,* 2009, pp. 85-97.

37 Consider the criticism towards the European Group on Ethics in Science and New Technologies (EGE) in the context of the EU by Helen Busby/Tamara Hervey/Alison Mohr, "Ethical EU Law? The Influence of the European Group on Ethics in Science and New Technologies", *European Law Review* 33 (2008), pp. 803-842.

38 Cf. e.g. Harvard Faculty of Medicine: Policy on Conflicts and Commitment (1990) appointing a Standing Committee on Conflicts of Interest and Commitment; European Federation of Psychologists Association: Recommendations on evaluative procedures and corrective actions in case of complaints about unethical conduct (2005).

of State sovereignty, international or transnational rules[39] are limited to the prohibition of the development – which, of course, includes development research – of biological weapons and restrictions upon the development of nuclear weapons. The prohibition of the development of biological weapons is at the core of the Convention on the Prohibition of the Development, Production and Stockpiling of Bacteriological (Biological) and Toxin Weapons and on Their Destruction (1972)[40] Concerning nuclear weapons, the relevant Treaty on the Non-Proliferation of Nuclear Weapons[41] does not contain an explicit prohibition of research, but implies that such research is restricted as it expressly facilitates nuclear research for peaceful purposes[42]. In the context of military research, the imperative of the peaceful use of the outer space (including research) is of certain importance[43].

---

[39] Sebastian Steinecke (supra note 4), manuscript pp. 374 *et seq.*

[40] 1015 U.N.T.S. 163.

Article I: "*Each State Party to this Convention undertakes never in any circumstances to develop, produce, stockpile or otherwise acquire or retain:*

(1) Microbial or other biological agents, or toxins whatever their origin or method of production, of types and in quantities that have no justification for prophylactic, protective or other peaceful purposes;

(2) Weapons, equipment or means of delivery designed to use such agents or toxins for hostile purposes or in armed conflict." (emphasis by the authors).

[41] 729 U.N.T.S. 161.

[42] Article IV (2): "*Nothing in this Treaty shall be interpreted as affecting the inalienable right of all the Parties to the Treaty to develop research, production and use of nuclear energy for peaceful purposes without discrimination and in conformity with Articles I and II of this Treaty.*"

[43] Cf. Article IV (2) of the Treaty on Principles Governing the Activities of States in the Exploration and Use of Outer Space, including the Moon and Other Celestial Bodies (U.N.T.S. 610, 205): "*The moon and other celestial bodies shall be used by all States Parties to the Treaty exclusively for peaceful purposes. The establishment of military bases, installations and fortifications, the testing of any type of weapons and the conduct of military manoeuvres on celestial bodies shall be forbidden. The use of military personnel for scientific research or for any other peaceful purposes shall not be prohibited. The use of any equipment or facility necessary for peaceful exploration of the moon and other celestial bodies shall also not be prohibited.*"

Another area of ethical relevance is the protection of animals in research[44].

3. The Promotion of Research and the Position of the Researcher

The third and last purpose for the governance of science is only marginally concerned with ethical issues. Its main impetus is the promotion of science by States and international entities and the rulemaking accompanying it. If science is supported (particularly by financial means) by such institutions, the structure, contents and aims of research may be planned and research may be directed and monitored. Substantive programmes of expenditure and scientific development need regulated structures. In this context, it may be considered necessary to preserve the independent – and sometimes also social – position of the researcher, in particular if research is undertaken in the framework of private institutions.

III. Rulemaking and Standard-Setting

1. Consensual Rulemaking

After identifying the most pertinent governance purposes, it is now necessary that we turn to the methods of rulemaking and standard-setting applied to realise these purposes at a normative level – before considering issues of implementation and management. In this respect, we have to acknowledge first, that in classical international law, consent lies at the basis of the creation of legal principles and rules[45]. The core instrument of consensual rulemaking is, of course, the international

---

44 Cf. an overview and link list of rules and guidelines in connection with animal research http://www.codex.uu.se/en/forskningdjur.shtml; cf., e.g., the Resolution on the Use of Animals in Research, Testing, and Education American Psychological Association (1990), available at: http://www.apa.org/science/leadership/care/animal-resolution.pdf, or the Statement on Animal Use in Biomedical Research of the World Medical Association (2006), available at http://www.wma.net/en/30publications/10policies/a18/index.html.

45 Cf. supra C. I.

treaty[46]. In many international organisations, the lawmaking function is exercised by the power to propose the conclusion of such treaties which are then submitted to the Member States for signature and ratification[47]. In the field of science, this applies in particular to the UNESCO. Article IV (B) (4) of its Constitution states:

> The General Conference shall, in adopting proposals for submission to the Member States, distinguish between recommendations and international conventions submitted for their approval. In the former case a majority vote shall suffice; in the latter case a two-thirds majority shall be required. Each of the Member States shall submit recommendations or conventions to its competent authorities within a period of one year from the close of the session of the General Conference at which they were adopted.

Thus, by a two-thirds majority, UNESCO may propose international treaties (conventions) to the member States for them to be submitted to ratification. This has not yet been done, though. The most important convention under the regime of UNESCO, the Convention Concerning the World Cultural and Natural Heritage[48] is related to matters outside the governance of science and the purposes developed here[49].

Finally, the conclusion of agreements not considered as international treaties must be considered in this context. Examples are above all cooperative agreements between sub-state entities (e.g. universities) and/or private actors (e.g. non-state research institutions)[50].

---

46 Cf. the seminal article by Christian Tietje, "The Changing Legal Structure of International Treaties as an Aspect of an Emerging Global Governance Architecture", *German Yearbook of International Law* 42 (2000), pp. 26-55.

47 A comprehensive survey and analysis is offered by José E. Alvarez (supra note 2), at pp. 273 *et seq.*

48 1037 U.N.T.S. 151.

49 Cf. only Diana Zacharias, "The UNESCO Regime for the Protection of World Heritage as Prototype of an Autonomy-Gaining International Institution", in: von Bogdandy et al. (supra note 2), pp. 301-336 at p. 301.

50 These actors are explained supra D. IV.

2. Institutional Rulemaking

a) Rulemaking by International Organisations

The role of International Organisations as rulemaking institutions at global (and regional) level has become obvious in recent times[51]. As already mentioned, the dichotomy between such rules that are legally binding and others that are devoid of such effect and deemed to be confined to political influence or convincing effects only is still considered to be of analytical value. Binding rules of the UN Security Council or certain specialised international agencies are opposed to recommendatory rules of other international institutions[52]. Before taking a deeper look into this aspect, it is important to note that in the field of science, International Organisations that issue rules *at global level* are indeed not empowered to endow them with strictly binding character.

This applies for UNESCO recommendations (cf. again Article IV (B) (4) of the UNESCO-Constitution, above). Such recommendations do not have binding force in a sense that they must be implemented by all means, but Member States are obliged to report on whether and in which way they have implemented them (Article VIII UNESCO-Constitution):

> *Reports by Member States*
>
> Each Member State shall submit to the Organization, at such times and in such manner as shall be determined by the General Conference, reports on the laws, regulations and statistics relating to its educational, scientific and cultural institutions and activities, and on the action taken upon the recommendations and conventions referred to in Article IV, paragraph 4.

In this context, the UNESCO has issued a series of recommendations to be implemented within the Member States. The Recommendation concerning the Status of Higher-Education Teaching Personnel (1997)[53]

---

[51] Cf. the seminal work by José E. Alvarez (supra note 2).

[52] For the Security Council cf. José E. Alvarez (supra note 2), pp. 199 *et seqq.*; Jurij Daniel Aston, *Sekundärgesetzgebung internationaler Organisationen zwischen mitgliedstaatlicher Souveränität und Gemeinschaftsdisziplin*, 2005, pp. 68 *et seqq.*; Stefan Talmon, "The Security Council as World Legislature", *The American Journal of International Law* 99 (2005), pp. 175-193; for other institutions cf. Aston, op. cit., pp. 126 *et seqq.*

[53] http://portal.unesco.org/en/ev.php-URL_ID=13144&URL_DO=DO_TOPIC&URL_SECTION=201.html.

contains detailed standards for the work and working conditions of (inter alia) teaching scientists.

Another means of rulemaking is the issuing of declarations. Such declarations are not strictly legally binding but may influence national policies in providing a standard of implementation in a certain policy field. Within such declarations, the duty of member States to implement them is usually laid down. Such declarations are promulgated e.g. by UNESCO, though not explicitly provided for by the UNESCO-Constitution. Nonetheless, the General Declaration on the Human Genome and Human Rights (1997)[54], the International Declaration on Human Genetic Data (2003)[55] and the Universal Declaration of Bioethics and Human Rights (2005)[56] contain some of the most important guidelines in international bioethical governance.

A related means of formulating general rules in a *prima facie* non-binding manner is the establishment of guidelines or similar instruments by subsidiary organs. This applies e.g. to the Ethical Guidelines for International Comparative Social Science Research in the framework of Management of Social Transformations (MOST) within UNESCO[57]. These guidelines are headed by the following overall objective explaining their *raison d'être* and giving an orientation towards their interpretation:

> "Research should be conducted in such a way that the integrity of the research enterprise is maintained, and negative after-effects which might diminish the potential for future research should be avoided."

---

[54] Available at http://www1.umn.edu/humanrts/instree/Udhrhg.htm. Cf. the Implementation: Noëlle Lenoir, "Universal Declaration on the Human Genome and Human Rights: The First Legal and Ethical Framework at the Global Level", *Columbia Human Rights Law Review* 30 (1998-1999), pp. 537-587; Thomas Alured Faunce, "Will international human rights subsume medical ethics? Intersections in the UNESCO Universal Bioethics Declaration", *Journal of Medical Ethics* 31 (2005), pp.173–178 at p. 174.

[55] http://portal.unesco.org/en/ev.php-URL_ID=17720&URL_DO=DO_TOPIC&URL_SECTION=201.html.

[56] Available at http://unesdoc.unesco.org/images/0014/001461/146180e.pdf.

[57] Available at http://www.unesco.org/most/ethical.htm.

A parallel example is provided by the OECD Principles and Guidelines for Access to Research Data from Public Funding[58]. These Principles and Guidelines

> "... are intended to promote data access and sharing among researchers, research institutions, and national research agencies, while at the same time, recognising and taking into account, the various national laws, research policies and organisational structures of member countries."

Simultaneously, the protection of intellectual property is considered and guaranteed according to domestic law[59].

b) Supranational Rulemaking

Among International Organisations, the toolbox for rulemaking vested in the EU is outstanding, which has lead to a discussion about whether supranational organisations can still be ranked among International Organisations[60]. At any rate, the entire range of legislative and rulemaking measures and procedures can be applied within the EU's research, technological development and space chapter. The framework programmes are passed as decisions in the ordinary legislative procedure (Articles 182 (1) and 289 (1), 294 TFEU)[61], specific programmes according to a special legislative procedure after consulting the European Parliament and the Economic and Social Committee (Articles 182 (4) and 289 (2) TFEU)[62]. That special legislative procedure also provides for the initiative of the Commission (Article 17 (2), first sentence TEU). The ordinary legislative procedure is further applicable for setting up joint undertakings and similar structures (Article 187, 188 (1) TFEU) as well as other supporting measures (Article 188 (2)), and finally for the European space policy (Article 189 (1) TFEU). Additional measures to implement the European Research Area may also be passed in the ordinary legislative procedure (Article 182 (5) TFEU). An important example is Regulation (EC) 1906/2006 for the technical implementation of

---

58 Available at http://www.oecd.org/document/2/0,3343,en_2649_34293_38500791_1_1_1_1,00.html.

59 (III) (E) of the Principles.

60 Cf. above D. III. 2. a.

61 Cf. Matthias Ruffert, in: Christian Calliess and id. (eds.), *EUV/AEUV*, 4th ed. 2011, Art. 182 TFEU, para. 5.

62 Cf. Matthias Ruffert (supra note 61) Art. 182 TFEU, para. 10.

the 7th Framework Programme[63]. Furthermore, delegated and implementing rules may be adopted (Articles 290 and 291 TFEU)[64], and, in co-operation with third countries, international agreements may by concluded (Articles 186 (2) and 216 TFEU).

These broad legislative powers notwithstanding, the EU applies – in the field of science and elsewhere – methods of non-binding rulemaking for an obvious reason: Within the institutional framework of the EU (supra D. III. 2. c.), the assent of the Council of the European Union is needed which assembles representatives of the Member States at ministerial level. Binding legislation is often considered as a powerful intrusion into the sphere of sovereignty of the Member States. Therefore, a comprehensive picture of the implementation of the EU's research policy also has to take into account non-legislative measures of governance, whether in the context of rulemaking or in the field of implementation and management.

An illustrative example of this mode of rulemaking governance is the European Charter for Researchers. This Charter is a formal recommendation of the Commission to the Member States[65]. As such, it is not binding in the clear sense that EU Law gives to this quality (Article 288 (5) TFEU)[66]; the Member States are just recommended to implement it within their own governance of research. If that was the only effect, the rather strong criticism against the Charter would be devoid of justification[67]. The Charter intends to lay down ethical scales for researchers

---

63 [2006] O.J. L 391/1.

64 On these types of EU legislation Herwig Hofmann, "Legislation, Delegation and Implementation under the Treaty of Lisbon: Typology Meets Reality", *European Law Journal* 15 (2009), pp. 482-505.

65 [2005] O.J. L 75/67.

66 Damian Chalmers/Gareth Davies/Giorgio Monti, *European Union Law,* 2nd ed. 2010, p. 101.

67 See the criticism by the German Bundesrat (BR Drs. 715/03) and the German Principals' Conference (Hochschulrektorenkonferenz), Beschluss vom 23 November 2005 (available at http://www.hrk.de/de/download/dateien/Empfehlung_Charta.pdf). Less critical is the assessment by Wissenschaftsrat, *Empfehlungen zur deutschen Wissenschaftspolitik im Europäischen Forschungsraum*, 2010 (Drucksache 9866-10), at pp. 79 *et seq.* and Alexander von Humboldt-Stiftung, Arbeitsgemeinschaft industrieller Forschungsvereinigungen "Otto von Guericke" e.V., Deutscher Akademischer Austauschdienst, Deutsche Forschungsgemeinschaft, Fraunhofer Gesellschaft, Helmholtz-Gemeinschaft Deutscher Forschungszentren, Hochschulrektorenkonferenz, Leibniz-Gemein-

and tries to optimise their working conditions. These purposes however imply considerable intrusions into the activities of researchers. Strangely enough, it even starts – although under the headline "Research Freedom" with an obligation:

> "Researchers should focus their research for the good of mankind and for expanding the frontiers of scientific knowledge, while enjoying the freedom of thought and expression, and the freedom to identify methods by which problems are solved, according to recognised ethical principles and practices."

The other provisions of the Charter also concentrate less on academic freedom but on the implementation of labour law requirements in researcher's employment situation and on the use of research for society at large. It is arguable that choosing the form of recommendation would allow Member States to set aside provisions which do not match their understanding of the freedom of research. This would however be short-sighted: The recommendation itself mentions monitoring procedures, and it is not excluded (to say the least) that it will be taken into account to establish criteria in EU funding mechanisms. It is obvious that rulemaking is strongly linked to implementation and management here.

c) Private Transnational Rulemaking

If the analysis of international science-related rulemaking was limited to State consent and State operated international or even supranational institutions, a considerable part of rules governing scientific activity would be omitted. As described in the context of actors, a considerable number of institutions that are totally (and in exceptions at least partly) independent of States and International Organisations (which are, by definition, founded by States as subjects of international law). The rulemaking activity of such private institutions largely influences research at a global scale. The most prominent example is provided by the

---

schaft, Stifterverband für die Deutsche Wissenschaft, Wissenschaftsrat (of 1 October 2006, available at: http://www.humboldt-foundation.de/pls/web/docs/F21999/2006_forschercharta_erklaerung.pdf). However less critical Henning Eikenberg, in: Eberhard Grabitz and Meinhard Hilf (eds.), *Das Recht der Europäischen Union, Band I EUV/EGV*, 2005, Preliminaries to Art.163-173 EGV, para. 41.

rules of the World Medical Association promulgated in its Helsinki Declaration which covers all scientific activity in the medical field[68].

According to the traditional concept of international legal rules, such declarations and other instruments issued by private institutions are not considered as valid sources. However, their guiding effect is by no means impaired by such categorisations. Medical research undertaken without respect for the Helsinki Declaration will be devoid of public funding. It may be considered as illegal – even as a criminal offence – if the standards of the Declaration are integrated into the interpretation of domestic legal rules[69].

IV. Implementation and Management

1. Implementation of Rules

As stated initially, rulemaking shall not be considered in an isolated way. Governance implies the implementation of rules; global administrative law encompasses sound administration. Specific methods for the implementation of international rules in science have already been mentioned: ethics committees, independent expert bodies and funding programmes, to name but a few.

2. Reporting and Benchmarking

The division between legal and social norms and the concomitant issue of the binding quality of rules put aside, what has been described up to this point including the issue of rule implementation could be summed up as governance by rules. The establishment of rules and their implementation is a core target of legal scrutiny, in a way that standards and procedures for control in legal terms are in existence or can be established following more or less traditional patterns.

Nonetheless, in recent times, international institutions have often used different tools of governance. The mere reporting about the development in certain States with respect to a particular policy field may de-

68 Cf. above E. II. 2.

69 Cf. Sebastian Steinecke, (supra note 4), manuscript pp. 383 *et seq.*

tect progress and deficiencies – from the point of view of defined policy aims of the institutions – and encourage or even factually compel the government of the relevant State to take action. The policy aims may be defined in a common benchmarking procedure[70].

To a great extent, the activities of the OECD in the field of science are operated alongside these mechanisms. The OECD Convention (1960) only contains scarce provisions carrying this process, and they are, of course, neither limited to nor specified at the field of science:

> Article 3
>
> With a view to achieving the aims set out in Article 1 and to fulfilling the undertakings contained in Article 2, the Members agree that they will:
>
> (a) keep each other informed and furnish the Organisation with the information necessary for the accomplishment of its tasks;
>
> (b) consult together on a continuing basis, carry out studies and participate in agreed projects; and
>
> (c) co-operate closely and where appropriate take co-ordinated action.

On this basis of common consultation and information, the OECD regularly issues reports on the development of science and technology in its Member States[71]. Statistical data is collected and published biannually (Main Science and Technology Indicators – MSTI)[72]. In the field of governance of public research, six country studies on the main characteristics of the public research systems and the recent reforms undertaken in each country have been carried out and proved to be rather influential[73]. Particular surveys have been undertaken with respect to tar-

---

70 Cf. generally Dirk Lehmkuhl, *Governance by Rating and Ranking*, Paper presented at the annual meeting of the International Studies Association, 2005, available at http://www.allacademic.com//meta/p_mla_apa_research_citation/0/7/0/7/3/pages70738/p70738-1.php.

71 Methodology OECD Country Notes for Science and Innovation (2008) – overview and link list available at http://www.kooperation-international.de/en/oecd/themes/info/detail/data/37052/?PHPSESSID=c332&cHash=fa1045e23b.

72 http://www.oecd.org/document/26/0,3343,en_2649_34273_1901082_1_1_1_1,00.html.

73 http://www.oecd.org/document/39/0,2340,en_2649_37417_2507879_1_1_1_37417,00.html.

geting research and development[74], respective tax incentives[75] and the financing of research and development[76]. A current project analyses new approaches and governance mechanisms for multilateral co-operations in science, technology and innovation to address global challenges[77].

Of course, if shortcomings are detected in such reports and studies, this promotes public debate within the Member State and is likely to bring forward domestic policy changes – effects that could never have been reached by any rulemaking effort of the organisation. When (re-)considering rulemaking and management in the field of science, such effects have to be taken into account to avoid the circumvention of legal caveats. It is also to be avoided that domestic representatives chose to implement policies via the OECD or other organisations that could not be implemented by way of the usual domestic procedures.

The EU has taken up this mode of governance in a very effective mechanism of governance referred to as co-ordination or sometimes "open method of co-ordination" (OMC)[78]. Instead of establishing legally binding norms to be respected and implemented by the Member States, the EU fixes benchmarks on the basis of voluntary consultations. Those benchmarks are not necessarily binding, but a continuous monitoring and reporting process reveals in regular terms which level of performance can be detected in the different Member States. The bare knowledge about success or bad performance in one Member State in relation to other Member States can create public pressure which may be at least as effective as binding legislation. It might be discussed

---

[74] Targeting R & D: Economic and Policy Implication of increasing R&D Spending (Working Paper, 2003), OECD Doc. DSTI/DOC(2003)8 of 24 July 2003.

[75] Tax Incentives for Research and Development: Trends and Issues, 2003, available at: http://www.oecd.org/dataoecd/12/27/2498389.pdf.

[76] Public and Private Financing of Business R&D, available at http://www.oecd.org/dataoecd/24/63/33719811.pdf.

[77] OECD-Project "New governance approaches for multilateral research to address global challenges" – http://www.die-gdi.de/CMS-Homepage/openweb cms3_e.nsf/%28ynDK_contentByKey%29/MSIN-7V3GZD?OpenDocument &nav=expand:Research%20and%20Consulting\Projects;active:Research%20a nd%20Consulting\Projects\MSIN-7V3GZD.

[78] Cf. only Sabrina Regent, "The Open Method of Coordination: A New Supranational Form of Governance?", *European Law Journal* 9 (2003), pp. 190-214.

that such influence upon legal and political systems of the Member States needed a specific treaty competence and cannot just be applied in shape of the "open method of co-ordination", but this argument is met by the Lisbon Treaty, as the relevant text of Article 181 provides that

> "(1) The Union and the Member States shall coordinate their research and technological development activities so as to ensure that national policies and Union policy are mutually consistent."

and that the Commission may take initiatives

> "(2)... aiming at the establishment of guidelines and indicators, the organisation of exchange of best practice, and the preparation of the necessary elements for periodic monitoring and evaluation. The European Parliament shall be kept fully informed."

Similar to the EU's employment policy, the coordination method explicitly applies to research and technological development.

3. Management by Contract

A final technique to achieve research goals is relating to a legal tool, but not to the establishment of general rules. In some instances, international organisations are concluding contracts with research institutions whether inside or outside universities. This is a particular method of research sponsorship within the European Union[79].

---

79 Matthias Ruffert (supra note 61), Art. 182, para. 8.

F. The Global Administrative Law of Science Revisited

I. The Global Governance of Science and Global Administrative Law

When undertaking to sum up the results of this study, two interrelated questions occur: (1) What can be inferred from the global governance of science in particular with respect to the development of global administrative law? (2) In which way do the theory and concept of global administrative law contribute to a sound global governance of science? Of course, it is these questions that underlie the study as a whole, but it is worth briefly revisiting the subject after now having gone through the constitutional basis, the institutional variety and the plenitude of governance mechanisms. In combining these issues of the foregoing chapters, there is first of all the question whether the institutional and normative abundance can be arranged satisfactorily (infra sub II.). Secondly, particularly in light of the constitutional orientation of freedom of science, it must be asked whether and how the institutional arrangements and the mechanisms of governance are legitimate and afford sufficient means of legal protection (infra sub III.).

II. From Sources to Rules and Standards

The unease with the traditional doctrine of public international legal sources has already been mentioned (supra E. I.). *Jan Klabbers* has recently illustrated it very strongly: "... upon closer inspection, sources doctrine has never failed to disappoint."[1] Thus, it should not be astonishing that the divergent types of rulemaking and standard-setting elaborated in this study (supra E. III.) can rarely be categorised in tradi-

[1] See Jan Klabbers, "Goldmann Variations", in: Armin von Bogdandy/Rüdiger Wolfrum/Jochen von Bernstorff/Philipp Dann/Matthias Goldmann (eds.), *The Exercise of Public Authority by International Institutions*, 2010, pp. 713-725 at p. 713.

tional patterns, let alone in the narrow framework of Article 38 ICJ-Statute.

Furthermore, scholars transcending the traditional positions have elaborated the lack of complexity, above all of the term "soft law", as its genetic, textual and follow-up parameters can reveal a broad range of instruments of different creation, shape and implementation[2]. This applies without any doubt for the normative global governance of science, as well. Effective standardisation, whether in the fields of ethical control or the promotion of science is quite rarely effectuated by means of classical international treaties, and actually traditional rulemaking by international organisations is not dominant – even in instances such as the European Union where the relevant international organisation is vested with broad and effective supranational rulemaking powers.

In such a regulatory atmosphere, the categorical distinction between "binding" and "non-binding" loses its overall importance in the first place. True, from a strong positivist viewpoint the whole matter would lose its position in legal scholarship at this point, but this would be short-sighted, and it is submitted that the brilliant positivists often quoted in this context would not get caught into the obvious trap of ignoring strong legal effects of standards that are not binding in the first place such as UNESCO-recommendations, the Helsinki Declaration or the European Charter for Researchers[3]. It is apparently insufficient to consider the last implementing (domestic) legal instrument only if the content of the rules and principles is legally analysed or even challenged. If we have a clear constitutional orientation such as freedom of science, intrusions and incursions by "non-binding" instruments can-

---

2 See only Alan E. Boyle, "Some Reflections on the Relationship of Traties and Soft Law", *International and Comparative Law Quarterly* 48 (1999), pp. 901-913; Christine M. Chinkin, "The Challenge of Soft Law: Development and Change in International Law", *International and Comparative Law Quarterly* 38 (1989), pp. 850-866. Cf. also the critical assessment by Prosper Weil, "Towards Relative Normativity in International Law?", *American Journal of International Law* 1983, pp. 413-442; and Jan Klabbers, "The Undesirability of Soft Law", *Nordic Journal of International Law* 67 (1998), pp. 381-391, as well as the assessment in terms of international relations by Kenneth W. Abbott and Duncan Snidal, "Hard and Soft Law in International Governance", *International Organization* 54 (2000), pp. 421-456.

3 Matthias Goldmann, "Inside Relative Normativity: From Sources to Standard Instruments for the Exercise of International Public Authority", in: Armin von Bogdandy et al. (supra note 1), pp. 661-711 at p. 672, footnote 47 quotes Austin, Kelsen, Hart and Luhmann.

not be ignored. Their categorisation has been undertaken from various viewpoints[4], and even if the efforts to achieve systematic structure and sound order here are far from being completed, there is no need to fear the disaggregation of public international law by the transformation of relative (but controlled) normativity into an amorphous plethora of norms with no clear arrangement of validity or even hierarchy.

Considering the variety of "soft" categories of norms, what is pertinent in the field of science is the prominence of standards generated by private or at least hybrid actors: networks of scientific institutions, professional bodies or other non-state actors (cf. supra D. IV., V.). Domestic administrative law has for a long time sought for appropriate terminological and conceptual tools for the integration of private standard-setting into its scope of application. This is mainly effectuated by concentrating upon the (domestic) act of recognition or incorporation of such rulemaking[5]. This strategy is far less viable in global administrative law, given that the elaboration of norms and their final implementation do regularly take place at different regulatory levels – international, transnational, supranational and domestic. If we are searching for mechanisms to assess the normative framework of the global governance of science that is globally operational, rulemaking and standard-setting activity involving private actors cannot be left apart.

III. From Effective Governance to Legitimate Administration

Moreover, in the field of science "soft" rulemaking even fades to implementation and management without rules, to standard-setting integrated within governance procedures without the prior distinct elaboration of rules. This is best illustrated in the subtleness of benchmarking- and ranking-procedures, e.g. inside the framework of the OECD (supra D. III. 1. d.). Obviously, nothing but very vague aims are fixed standards in the beginning of such (co-ordinating) procedures, but the

---

4 See above all Matthias Goldmann (supra note 3), at pp. 684 *et seq.*

5 Cf. Matthias Ruffert, "Rechtsquellen und Rechtsschichten des Verwaltungsrechts", in: Wolfgang Hoffmann-Riem/Eberhard Schmidt-Aßmann/Andreas Voßkuhle (eds.), *Grundlagen des Verwaltungsrechts, Band I,* 2006, paras. 17/18 *et seq.* and 85 *et seq.* The issue of legitimacy in this context is generally treated by Hans-Georg Dederer, *Korporative Staatsgewalt,* 2004.

core and precise standards to be implemented are elaborated within the "ping-pong-procedure" of benchmarking and control. It is submitted that particularly in light of the free performance of science such procedures which are norm-creating and norm-implementing but not rule-making require the same safeguards in terms of legitimacy and legal protection[6].

Introducing the term legitimacy at this stage bears the risk of transforming a rather clear-cut study into a never-ending story. Fortunately enough, it is possible to shorten this issue by underlining the two core limbs of legitimacy in public law: democracy and bureaucracy, the former being indispensable (and not replaceable, above all not by the latter) in modern constitutionalism, the latter integrating elements of efficacy and good governance to shape viable administrative structures[7]. Consequently, any mechanism that does not have a procedural basis that could be designated as democratic must be rejected and any mechanism that is not bureaucratically effective should be discarded and replaced by a more suitable one. What has to be avoided in terms of democratic rule is the circumvention of domestic democratic (parliamentary) procedure by national executives via rulemaking-mechanisms of global governance[8].

In this respect, the broad integration of rule-making that cannot be traced back to any State authority represents a considerable problem. The categorical distinction between private freedom on the one hand and the restrictions upon public authority on the other hand must be upheld on the global scale. Restrictions on private rulemaking are un-

---

6 Cf. the assessment by Matthias Knauff, *Der Regelungsverbund: Recht und Soft Law im Mehrebenensystem*, 2010, at pp. 459 *et seq*. This is opposed to the view of Jean d'Aspremont, "Softness in International Law: A Self-Serving Quest for New Legal Materials", *European Journal of International Law* 19 (2008), pp. 1075- 1093, who intends to exclude the analysis of soft law from international law and underlines the importance of non-legal developments.

7 Cf. Stefan Kadelbach, "Demokratische Legitimation als Prinzip zwischenstaatlichen Handelns", in: Silja Vöneky/Cornelia Hagedorn/Miriam Clados/Jelena von Achenbach (eds.), *Legitimation ethischer Entscheidungen im Recht,* 2009, pp. 147-171; José E. Alvarez, *International Organisations as Lawmakers,* 2005, at pp. 627 *et seq.*; Steve Charnovitz, "The Emergence of Democratic Participation in Global Governance", *Indiana Journal of Global Legal Studies* 10 (2003), pp. 45-77.

8 See Kerstin Marten and Klaus Dieter Wolf, "Paradoxien der Neuen Staatsräson", *Zeitschrift für Internationale Beziehungen* 13 (2006), pp. 145-176 at p. 148, and Matthias Goldmann (supra note3), at p. 668.

dergoing the risk of implementing limitations on the free associative will of private individuals. Therefore, it is impossible to confine private standard-setting, management and implementation activities to the same boundaries as State-generated governance. Thus it is necessary to distinguish between freedom of science as a limit to public governance in the global sphere and the duty of publicly founded governance institutions to preserve freedom of science in the various institutional frameworks against intrusions from private or hybrid actors. Further, the more global governance in the field of science is transferred to non-State actors such as professional bodies or other NGOs, the more there is a responsibility within international organisations and domestic jurisdictions to safeguard freedom of science and to provide adequate legal protection to scientists and other bearers of that right (e.g. free research institutions).

IV. Conclusive Remarks

Indeed, the consequences the perpetrator had to face in the initial illustrative example were legal in any respect. Strict domestic rules continue to provide for the punishment of criminal activity around overt misbehaviour in the field of science. However, as could be seen in the present study, the scope of the "legal" in the international governance of science is far more insecure than in the initial case. Obviously, institutional patterns and regulatory mechanisms beyond what has hitherto been considered as part of international law along the division between binding and non-binding are particularly frequent in the governance of science. Obviously, this is partly due to the opposition or even hostility of science towards the law (cf. above A. IV.). This, however, cannot prevent the international (administrative) lawyer from considering the legal content of impact of institutions, principles and rules that are traditionally thought to be outside the scope of legal scrutiny and scholarship. The main outcome of this reflection is the informed awareness of possible intrusions of legal regulation within the free exercise of science, whether such intrusions are motivated by the protection of values outside science or whether they are even meant to promote science itself.

Science is impossible without a legally guaranteed freedom to exercise it. Elements of such freedom can be drawn from an analysis of international and domestic provisions and principles. Institutions and rules are to be constructed and interpreted around and in light of this freedom to guarantee the continuous existence of the knowledge-based society by means of a global administrative law of science.

Literature

Abbott, Kenneth W. and Duncan Snidal, "Hard and Soft Law in International Governance", *International Organization* 54 (2000), pp. 421 - 456

Alo, Babajide, "University-Based Applied Research and Innovation in Nigeria", in: Osita Ogbu/Banji Oyeyinka/Hasa Mlawa (eds.), *Technology Policy and Practice in Africa,* International Development Research Centre, Ottawa 1995, pp. 238 - 248

Alquié, Ferdinand, *Wissenschaft und Metaphysik bei Descartes,* Königshausen & Neumann, Würzburg 2001

Alvarez, José E., *International Organisations as Law-makers,* Oxford University Press, Oxford 2005

Amerasinghe, Chittharanjan Felix, "The Future of International Administrative Law", *The International and Comparative Law Quarterly* 45 (1996), pp. 773 - 795

Ancarani, Vittorio, "Globalizing the World – Science and Technology in International Relations", in: Sheila Jasanoff/Gerald E. Markle/James C. Petersen/Trevor Pinch (eds.), *Handbook of Science and Technology Studies*, Sage, Thousand Oaks (Calif.) 2005, pp. 652 - 670

Aquinas, Thomas, *The Disputed Questions On Truth* (Vol. I), translated by Robert William Mulligan, S. J. Chicago: Henry Regnery Co., 1952

Arsanjani, Mahnoush H., "The Negotiations on a Treaty on Cloning: Some Reflections", in: Silja Vöneky and Rüdiger Wolfrum (eds.), *Human Dignity and Human Cloning,* Nijhoff, Leiden 2004, pp. 145 - 165

Aston, Jurij Daniel, *Sekundärgesetzgebung internationaler Organisationen zwischen mitgliedstaatlicher Souveränität und Gemeinschaftsdisziplin*, Duncker & Humboldt, Berlin 2005

Aubrey, Henry G., *Atlantic Economic Cooperation: The Case of the OECD*, Praeger, New York 1967

Auby, Jean-Bernard, *La globalisation, le droit et l'État,* Montchrestien, Paris 2003

Barkhoff, Jürgen and Helmut Eberhart (eds.), *Networking across Borders and Frontiers*, Lang, Frankfurt am Main 2009

Barnes, S. Barry and R.G.A. Dolby, "The Scientific Ethos: A Deviant Viewpoint", *Archives Européennes de Sociologie* XI (1970), pp. 3 - 25

Bell, Daniel, *The coming of post-industrial society: A venture in social forecasting*, Basic Books, New York 1973

Berger, Michael, *Vertraglich nicht vorgesehene Einrichtungen des Gemeinschaftsrechts mit eigener Rechtspersönlichkeit*, Nomos, Baden-Baden 1999

Biaggini, Giovanni, "Die Entwicklung eines Internationalen Verwaltungsrechts als Aufgabe der Rechtswissenschaft", *Veröffentlichungen der Vereinigung der Deutschen Staatsrechtslehrer* 67 (2008), pp. 413 - 445

Bischoff, Joachim, *Mythen der New Economy. Zur politischen Ökonomie der Wissensgesellschaft,* VSA-Verlag, Hamburg 2001

Bogdandy, Armin von/Philipp Dann/Matthias Goldmann, "Developing the Publicness of Public International Law: Towards a Legal Framework for Global Governance Activities", in: Armin von Bogdandy/Rüdiger Wolfrum/Jochen von Bernstorff/Philipp Dann/Matthias Goldmann (eds.), *The Exercise of Public Authority by International Institutions*, Springer, Heidelberg 2010, pp. 3 - 32

Bogdandy, Armin von and Dietrich Westphal, "Der rechtliche Rahmen eines autonomen Europäischen Wissenschaftsrates", *Wissenschaftsrecht* 37 (2004), pp. 224 - 255

Bok, Bart and Lawrence Jerome (eds.), *Objections to Astrology*, Prometheus Books, Buffalo 1975

Boyle, Alan E., "Some Reflections on the Relationship of Treaties and Soft Law", *International and Comparative Law Quarterly* 48 (1999), pp. 901 - 913

Breining-Kaufmann, Christine, "Internationales Verwaltungsrecht", *Zeitschrift für Schweizerisches Recht* 125 (2006), II, pp. 5 - 73

Brownlie, Ian, *Principles of Public International Law*, Oxford University Press, Oxford, 7th ed. 2008

Brudermüller, Gerd, "Ethikkommissionen und ethischer Diskurs", in: id. (ed.), *Angewandte Ethik und Medizin*, Königshausen & Neumann, Würzburg 1999, pp. 85 - 116

Busby, Helen/Tamara Hervey/Alison Mohr, "Ethical EU Law? The Influence of the European Group on Ethics in Science and New Technologies", *European Law Review* 33 (2008), pp. 803 - 842

Byrne, J. Peter, "Academic Freedom. A Special Concern of the First Amendment", *Yale Law Review* 99 (1989), pp. 251 - 339

Byrne, J. Peter, "Constitutional Academic Freedom after Grutter. Getting Real About the 'Four Freedoms' Of A University", *University of Colorado Law Review* 77 (2006), pp. 929 - 953

Calliess, Christian, "The Charter of Fundamental Rights of the European Union", in: Dirk Ehlers (ed.), *European Fundamental Rights and Freedoms*, de Gruyter, Berlin 2007, pp. 518 - 539

Callon, Scott, *Divided sun: MITI and the breakdown of Japanese high-tech industrial policy, 1975-1993*, Stanford University Press, Stanford 1995

Cassese, Antonio, *International Law*, Oxford University Press, Oxford, 2nd ed. 2005

Cassese, Sabino, "Is There a Global Administrative Law?", in: Armin von Bogdandy/Rüdiger Wolfrum/Jochen von Bernstorff/Philipp Dann/Matthias Goldmann (eds.), *The Exercise of Public Authority by International Institutions*, Springer, Heidelberg 2010, pp. 761 - 776

Cassese, Sabino, "Administrative Law without the State? The Challenge of Global Regulation", *New York University Journal of International Law and Politics* 37 (2005), pp. 663 - 694

Cassese, Sabino, "European Administrative Proceedings", *Law and Contemporary Problems* 68 (2004), pp. 21 - 35

Castells, Manuel, *The Rise of the Network Society. The Information Age: Economic Society and Culture*, Vol. 1, Blackwell, Cambridge (Mass.) 1996

Chalmers, Alan Francis, *Science and its Fabrication,* University of Minneapolis Press, Minneapolis 1990

Chalmers, Alan Francis, *What is this thing Called Sciences,* University of Queensland Press, Hackett, 3rd ed. 1999

Chalmers, Damian/Gareth Davies/Giorgio Monti, *European Union Law,* Cambridge University Press, Cambridge, 2nd ed. 2010

Charnovitz, Steve, "The Emergence of Democratic Participation in Global Governance", *Indiana Journal of Global Legal Studies* 10 (2003), pp. 45 - 77

Chernow, Barbara (ed.), "Art. Academic Freedom", *The Columbia Encyclopedia*, Columbia University Press, Boston, 5th ed. 1993

Chinkin, Christine M., "The Challenge of Soft Law: Development and Change in International Law", *International and Comparative Law Quarterly* 38 (1989), pp. 850 - 866

Chiti, Edoardo, "The Emergence of a Community Administration: the Case of European Agencies", *Common Market Law Review* 37 (2000), pp. 309 - 334

Chiti, Edoardo, "Towards a Model of Independent Exercise of Community Functions?", in: Roberto Caranta/Mads Andenæs/Duncan Fairgrieve (eds.), *Independent Administrative Authorities*, British Institute of International and Comparative Law, London 2004, pp. 205 - 224

Classen, Claus-Dieter, "Die Entwicklung eines Internationalen Verwaltungsrechts als Aufgabe der Rechtswissenschaft", *Veröffentlichungen der Vereinigung der Deutschen Staatsrechtslehrer* 67 (2008), pp. 365 - 412

Claude, Richard, "Scientists' Rights and the Human Right to the Benefits of Science", in: Audrey Chapman (ed.), *Core obligations: Building a framework for political, social and cultural rights*, Intersentia, Antwerp/Oxford/New York 2002, pp. 247 - 278

Commission on Global Governance, *Our Global Neighbourhood*, Oxford University Press, Oxford 1995

Corkin, Joseph, "Science, Legitimacy and the Law: Regulating Risk Regulation Judiciously in the European Community", *European Law Review* 33 (2008), pp. 359 - 384

Craig, Paul, *EU Administrative Law*, Oxford University Press, Oxford 2006

Craig, Paul, "The Constitutionalization of Community Administration", *Jean Monnet Working Paper* 3/2003, available at http://centers.law.nyu.edu/jeanmonnet/papers/03/030301.pdf

Crivelli, Paolo, *Aristotle on Truth*, Cambridge University Press, Cambridge 2006

Cronin-Furman, Kathleen Renée, "60 Years of the Universal Declaration of Human Rights: Towards an Individual Responsibility to

Protect", *The American University International Law Review* Vol. 25 Issue 1 (2009), pp. 175 - 198

Cyranoski, David, "Woo Suk Hwang convicted, but not of fraud", *Nature* 461 (27 October 2009), p. 1181

d'Aspremont, Jean, "Softness in International Law: A Self-Serving Quest for New Legal Materials", *European Journal of International Law* 19 (2008), pp. 1075 - 1093

Dederer, Hans-Georg, *Korporative Staatsgewalt*, Mohr Siebeck, Tübingen 2004

Demuro, Ginamario, "Article 13", in: William B.T. Mock/id./Raffaele Bifulco/Marta Cartabia/Alfonso Celotto (eds.), *Human Rights in Europe,* Carolina Academic Press, Durham 2010, pp. 84 - 87

Deutsch, Erwin and Jochen Taupitz, "Einführung: Forschungsfreiheit und Forschungskontrolle in der Medizin – zur geplanten Revision der Deklaration von Helsinki", in: id. (eds.), *Freedom and Control of Biomedical Research – The planned Revision of the Declaration of Helsinki*, Springer, Heidelberg 2000, pp. 1 - 6

Döhler, Elmar and Carsten Nemitz, "Wissenschaft und Wissenschaftsfreiheit in internationalen Vereinbarungen", in: Hellmut Wagner (ed.), *Rechtliche Rahmenbedingungen für Wissenschaft und Forschung. Forschungsfreiheit und Staatliche Regulierung*, Vol. 1, Nomos, Baden-Baden 2000, pp. 159 - 188

Drucker, Peter, *The age of discontinuity: guidelines to our changing society*, Hamper & Row, New York 1969

Dupuy, Pierre-Marie, "The Constitutional Dimension of the Charter of the United Nations Revisited", *Max Planck United Nations Yearbook* 1 (1997), pp. 1 - 33

Dürr, Hans-Peter, *Das Netz des Physikers,* Hanser, München 1988

Ehlers, Dirk, "§ 4 Europäisches Recht und Verwaltungsrecht", in: Hans-Uwe Erichsen and id. (eds.), *Allgemeines Verwaltungsrecht*, de Gruyter, Berlin, 14th ed. 2010, pp. 163 - 199

Eikenberg, Henning, "Preliminaries to Art.163-173", in: Eberhard Grabitz and Meinhard Hilf (eds.), *Das Recht der Europäischen Union, Band I EUV/EGV*, Beck, München 2005

Eikenberg, Henning, "Der Europäische Forschungsraum: Ein Kompetenzproblem?", in: *Europarecht* 2008, pp. 125 - 139

Esty, Daniel C., "Good Governance at the Supranational Scale: Globalizing Administrative Law, *Yale Law Journal* 115 (2006), pp. 1490 - 1562

Fassbender, Bardo, "The United Nations Charter As Constitution of the International Community", *Columbia Journal of Transnational Law* 36 (1998), pp. 529 - 619

Fassbender, Bardo, "Der Schutz der Menschenrechte als zentraler Inhalt des völkerrechtlichen Gemeinwohls", *Europäische Grundrechte-Zeitschrift* 30 (2003), pp. 1 - 15

Faunce, Thomas Alured, "Will international human rights subsume medical ethics? Intersections in the UNESCO Universal Bioethics Declaration", *Journal of Medical Ethics* 31 (2005), pp. 173 - 178

Fehling, Michael, "Article 5 (3)", in: Rudolf Dolzer/Karin Graßhof/Wolfgang Kahl/Christian Waldhoff (eds.), *Bonner Kommentar zum Grundgesetz*, Heidelberg, 148th ed. 2010

Fischer-Lescano, Andreas, "Transnationales Verwaltungsrecht", *Juristenzeitung* 2008, pp. 373 - 383

Franck, Thomas, "Is the UN Charter a Constitution?", in: Jochen Abr. Frowein (ed.), *Verhandeln für den Frieden, Liber Amicorum Tono Eitel*, Springer, Berlin/New York 2003, pp. 95 - 106

Frankfort, Henri/Henriette Frankfort, "Myth and Reality", in: id. et al. (eds.), *Before Philosophy. The Intellectual Adventure of Ancient Man*, Penguin Books, Harmondsworth (Middlesex) 1946, pp. 11 - 36

Freeman Jalet, Frances T., "The Quest for the General Principles of Law Recognized by Civilized Nations – a Study", *UCLA Law Review* 10 (1962-1963), pp. 1041-1086

Friedmann, Wolfgang, *The Changing Structure of International Law,* Stevens, London 1964

Froehlich, Annette, "Die European Space Policy und ihre Auswirkungen auf die Verlängerung des ESA/EG-Rahmenabkommens", *Zeitschrift für Luft- und Weltraumrecht* 57 (2008), pp. 67 - 77

Frowein, Jochen Abr. and Nico Krisch, in: Bruno Simma (ed.), *The Charter of the United Nations. A Commentary, Volume I,* Beck, München, 2nd ed. 2002

Gerken, Lüder, *Der Wettbewerb der Staaten,* Mohr Siebeck, Tübingen 1999

Glaser, Markus A., *Internationale Verwaltungsbeziehungen,* Mohr Siebeck, Tübingen 2010

Godt, Christine, "Forschungs-, Wissenschafts- und Technologiepolitik", in: Manfred Dauses, *Handbuch des EU-Wirtschaftsrechts,* Beck, München 2009, pp. 1 - 28

Goldberg, Steven, "The Constitutional Status of American Science", *University of Illinois Law Forum* 1 (1979), pp. 1 - 6

Goldmann, Matthias, "Inside Relative Normativity: From Sources to Standard Instruments for the Exercise of International Public Authority", in: Armin von Bogdandy/Rüdiger Wolfrum/Jochen von Bernstorff/Philipp Dann/Matthias Goldmann (eds.), *The Exercise of Public Authority by International Institutions,* Springer, Heidelberg 2010, pp. 661 - 711

Green, Maria, *Drafting History of the Artie 15 (1) (c) of the International Covenant on Economic, Social and Cultural Rights,* Background Paper, UN Doc. E/C.12/2000/15 (9 October 2000)

Grimm, Dieter, *Wissenschaftsfreiheit vor neuen Grenzen?,* Wallstein-Verlag, Göttingen 2007

Groß, Thomas, *Die Autonomie der Wissenschaft im europäischen Rechtsvergleich,* Nomos, Baden-Baden 1992

Hahn, Hugo J. and Albrecht Weber, *Die OECD: Organisation für Wirtschaftliche Zusammenarbeit und Entwicklung*, Nomos, Baden-Baden 1976

Hahn, Hugo J., "Organisation for Economic Co-operation and Development", in: Rudolf Bernhardt (ed.), *Encyclopedia of Public International Law*, Vol. 3, North-Holland Pub. Co., Amsterdam 1997, pp. 790 - 799

Heidenreich, Martin, "Die Debatte um die Wissensgesellschaft", in: Stefan Böschen and Ingo Schulz-Schaeffer (eds.), *Wissenschaft in der Wissensgesellschaft,* Westdeutscher Verlag, Wiesbaden 2003, pp. 25 - 51

Hermes, Georg, "Legitimationsprobleme unabhängiger Behörden", in: Hartmut Bauer/Peter M. Huber/Karl-Peter Sommermann (eds.), *Demokratie in Europa*, Mohr Siebeck, Tübingen 2005, pp. 457 - 490

Herriman, Michael, "Academic Freedom in Australia", *Interchange* 14 (1983), pp. 82 - 93

Higgins, John, "Academic Freedom in the New South Africa", in: *Boundary 2: International Journal of Literature and Culture* 27 (2000), pp. 97 - 119

Hilf, Juliane, "Preliminaries to Art. 163-173", in: Hans von der Groeben and Jürgen Schwarze (eds.), *Kommentar zum Vertrag über die Europäische Union und zur Gründung der Europäischen Gemeinschaft*, Nomos, Baden-Baden, 6th ed. 2003

Hilpert, Ulrich, "The State, Science and Techno-Industrial Innovation. A New Model of State Policy and a Changing Role of the State", in: id. (ed.), *State Policies and Techno-Industrial Revolution,* Routledge, London 1991, pp. 3 - 40

Hobe, Stephan/Katharina Kunzmann/Thomas Reuter/Julia Neumann, *Rechtliche Rahmenbedingungen einer künftigen kohärenten Struktur der Europäischen Raumfahrt*, LIT-Verlag, Berlin 2005

Hofmann, Herwig C. H. and Alexander Türk, "The Development of Integrated Administration in the EU and its Consequences", *European Law Journal* 13 (2007), pp. 253 - 271

Hofmann, Herwig, "Legislation, Delegation and Implementation under the Treaty of Lisbon: Typology Meets Reality", *European Law Journal* 15 (2009), pp. 482 - 505

Holden, Constance and Gretchen Vogel, "Cell Biology: A Seismic Shift for Stem Cell Research", *Science* 319 (2008), pp. 560 - 563

Hollinger, David, "The Defense of Democracy and Robert K. Merton's Formulation of the Scientific Ethos", *Knowledge and Society: Studies in the Sociology of Culture Past and Present* 4 (1983), pp. 1 - 15

Hüfner, Klaus and Wolfgang Reuther (eds.), UNESCO-Handbuch, Uno-Verlag, Bonn, 2nd ed. 2005

Human Rights Watch (ed.), *Academic Freedom in Indonesia, Dismantling Soeharto-Era Barriers,* 1998, available at http://www.hrw.org/legacy/reports98/indonesia2/

Human Rights Watch (ed.), *Reading Between the "Red Lines". The Repression of Academic Freedom in Egyptian Universities,* 2005, available at http://www.hrw.org/en/reports/2005/06/08/reading-between-red-lines-repression-academic-freedom-egyptian-universities

Humboldt, Wilhelm von, *Über die innere und äußere Organisation der höheren wissenschaftlichen Anstalten in Berlin* (1809/10), quoted in: Ernst Anrich (ed.), *Die Idee der deutschen Universität*, Genter, Darmstadt 1956, pp. 375 - 386

Hussain, Abid, *Civil and Political Rights Including the Question of Freedom of Expression*, E/CN.4/2000/63 of 18 January 2000

Hwang Woo-Suk et al., "Evidence of a Pluripotent Human Embryonic Stem Cell Line Derived from a Cloned Blastocyst", *Science* 303 (2004), pp. 1669 - 1674

Hwang Woo-Suk et al., "Patient-Specific Embryonic Stem Cells Derived from Human SCNT Blastocysts", *Science* 308 (2005), pp. 1777 - 1783

Izaguirre, Ines, "Argentina", *Academe* Vol. 85 (1999), pp. 18 - 20

Jessup, Philip C., *Transnational Law,* Yale University Press, New Haven 1956

Kadelbach, Stefan and Thomas Kleinlein, "International Law – a Constitution of Mankind? An Attempt at a Re-appraisal with an Analysis of Constitutional Principles", *German Yearbook of International Law* 50 (2007), pp. 303 - 347

Kadelbach, Stefan, "Demokratische Legitimation als Prinzip zwischenstaatlichen Handelns", in: Silja Vöneky/Cornelia Hagedorn/Miriam Clados/Jelena von Achenbach (eds.), *Legitimation ethischer Entscheidungen im Recht,* Springer, Heidelberg 2009, pp. 147 - 171

Kahl, Arno, "Europäische Agenturen im Lichte der dynamischen Verwaltungslehre", in: Konrad Arnold/Friederike Bundschuh-Rieseneder/id./Thomas Müller/Klaus Wallnöfer (eds.), *Recht Politik Wirtschaft Dynamische Perspektiven – Festschrift für Norbert Wimmer*, Springer, Wien 2008, pp. 245 - 284

Kakuk, Péter, "The Legacy of the Hwang Case: Research Misconduct in Biosciences", *Science and Engineering Ethics* 15 (2009), pp. 545 - 562

Kant, Immunael, "The Contest of Faculties", in: Hans Siegbert Reiss (ed.), *Kant: Political Writings*, Cambridge University Press, Cambridge, 2nd ed. 1991, pp. 176 - 190

Kayrooz, Carol/Pamela Kinnear/Paul Preston, "Academic Freedom and Commercialisation of Australian Universities: Perceptions and Experiences of Social Scientists", *Australia Institute Discussion Paper* No. 37 (2001)

Kingsbury, Benedict/Nico Krisch/Richard B. Stewart, "The Emergence of Global Administrative Law, *Law and Contemporary Problems* 68 (2005), pp. 15 - 61

Klabbers, Jan, "Constitutionalism Lite", *International Organization Law Review* 1 (2004), pp. 31 - 58

Klabbers, Jan, "Goldmann Variations", in: Armin von Bogdandy/Rüdiger Wolfrum/Jochen von Bernstorff/Philipp Dann/Matthias Goldmann (eds.), *The Exercise of Public Authority by International Institutions*, Springer, Heidelberg 2010, pp. 713 - 725

Klabbers, Jan, "The Undesirability of Soft Law", *Nordic Journal of International Law* 67 (1998), pp. 381 - 391

Kleindiek, Ralf, *Wissenschaft und Freiheit in der Risikogesellschaft*, Duncker & Humboldt, Berlin 1998

Knauff, Matthias, *Der Regelungsverbund: Recht und Soft Law im Mehrebenensystem*, Mohr Siebeck, Tübingen 2010

Knorr-Cetina, Karin, *The Manufacture of Science*, Pergamon Press, Oxford 1981

Kooiman, Jan (ed.), *Modern Governance: New Government-Society Interactions*, Sage, London 1993

Krajewski, Markus, *Verfassungsperspektiven und Legitimation des Rechts der Welthandelsorganisation*, Duncker & Humboldt, Berlin 2001

Kreibich, Rolf, *Die Wissenschaftsgesellschaft*, Suhrkamp, Frankfurt am Main, 2nd ed. 1986

Krisch, Nico, "The Pluralism of Global Administrative Law", *European Journal of International Law* 17 (2006), pp. 247 - 278

Kuhn, Thomas, *The Structure of Scientific Revolutions*, University of Chicago Press, Chicago, 3rd ed. 1996

Lachs, Manfred, "Thoughts on Science, Technology and World Law", *American Journal of International Law* 86 (1992), pp. 673 - 699

Lane, Robert, "The decline of politics and ideology in a knowledgeable society", *American sociological review* 31 (1966), pp. 649 - 662

Lehmkuhl, Dirk, *Governance by Rating and Ranking*, Paper presented at the annual meeting of the International Studies Association, 2005, available at http://www.allacademic.com//meta/p_mla_apa_research_citation/0/7/0/7/3/pages70738/p70738-1.php

Lenoir, Noëlle, "Universal Declaration on the Human Genome and Human Rights: The First Legal and Ethical Framework at the

Global Level", *Columbia Human Rights Law Review* 30 (1998-1999), pp. 537 - 587

Levin-Stankevich, Brian and Alexander Savelyev, "The Academic Profession in Russia", in: Philip G. Altbach (ed.), *The International Academic Profession. Portraits of Fourteen Countries,* Carnegie Foundation for the Advancement of Teaching, Princeton 1996, pp. 569 - 614

Lilie, Hans, "International Legal Limits to Human Cloning", in: Silja Vöneky and Rüdiger Wolfrum (eds.), *Human Dignity and Human Cloning,* Nijhoff, Leiden 2004, pp. 125 - 132

Lindberg, David, *Die Anfänge des abendländischen Wissens,* dtv, München 2000

Lindner, Josef Franz, *Die Europäisierung des Wissenschaftsrechts,* Mohr Siebeck, Tübingen 2009

Macdonald, Ronald St. John, "The United Nations Charter: Constitution or Contract?", in: id. and Douglas M. Johnston (eds.), *The Structure and Process of International Law,* Nijhoff, The Hague 1983, pp. 889 - 912

Macdonald, Ronald St. John, "The Charter of the United Nations in Constitutional Perspective", *Australian Yearbook of International Law* 20 (1999), pp. 205 - 231

Mann, Thomas, "Forschungsfreiheit und akademische Freiheit", in: F. Sebastian M. Heselhaus and Carsten Nowak (eds.), *Handbuch der Europäischen Grundrechte*, Beck/München, Linde/Wien and Stämpfli/Bern, 2006, pp. 745 - 770

Marten, Kerstin and Klaus Dieter Wolf, "Paradoxien der Neuen Staatsräson", *Zeitschrift für Internationale Beziehungen,* 13 (2006), pp. 145 - 176

Mayer, Franz, "Internationalisierung des Verwaltungsrechts?", in: Christoph Möllers/Andreas Voßkuhle/Christian Walter (eds.), *Internationales Verwaltungsrecht,* Mohr Siebeck, Tübingen 2007, pp. 49 - 72

Meessen, Karl M., "Souveränität im Wettbewerb der Systeme", in: Volkmar Götz/Peter Selmer/Rüdiger Wolfrum (eds.), *Liber Amicorum Günther Jaenicke - Zum 85. Geburtstag*, Springer, Berlin 1998, pp. 667 - 681

Mellein, Christine, "Art. 163", in: Jürgen Schwarze (ed.), *EU-Kommentar*, Nomos, Baden-Baden, 2nd ed. 2009

Merton, Robert K., *Social theory and social structure*, The Free Press, New York 1968

Merton, Robert K., "Science and Technology in a Democratic Order", *Journal of Legal and Political Sociology* 1 (1942), pp. 115 - 126

Merton, Robert K., "The Normative Structure of Science", in: Norman Storer (ed.), *The Sociology of Science: Theoretical and Empirical Investigations,* University of Chicago Press, Chicago 1973, pp. 267 - 280

Metzger, Walter, "The German Contribution to the American Theory of Academic Freedom", *American Association of University Professors Bulletin* 41 (1955), pp. 214 - 230, printed in: id. (ed.), *The American Concept of Academic Freedom in Formation. A Collection of Essays and Reports*, Arno Press, New York 1977

Metzger, Walter, "Profession and Constitution: Two Definitions of Academic Freedom in America", *Texas Law Review* 66 (1988), pp. 1265 - 1321

Möllers, Christoph, "Transnationale Behördenkooperation", *Zeitschrift für ausländisches öffentliches Recht und Völkerrecht* 65 (2005), pp. 351 - 389

Möllers, Christoph/Andreas Voßkuhle/Christian Walter (eds.), *Internationales Verwaltungsrecht,* Mohr Siebeck, Tübingen 2007

Möllers, Christoph, *Gewaltengliederung,* Mohr Siebeck, Tübingen 2005

Möllers, Christoph, "Pouvoir Constituant – Constitution – Constitutionalisation", in: Armin von Bogdandy and Jürgen Bast (eds.), *Principles of European Constitutional Law,* Hart, Portland, 2nd ed. 2010, pp. 183 - 226

Moriya, Kenichi, "Wissenschaftsfreiheit. Beobachtungen zum deutschen und japanischen juristischen Diskurs", *Rechtsgeschichte* 7 (2005), pp. 74 - 85

Mosler, Hermann, The International Society as a Legal Community, *Recueil des Cours* Vol. 140 (1974-IV), Sijthoff, Leiden 1974

Nemori, Ken, "Grenzen der Wissenschafts- und Forschungsfreiheit in der japanischen Verfassung und das Klonen von Menschen", in: Rainer Wahl (ed.), *Verfassungsrecht der Humangenetik im deutsch-japanischen Vergleich*, 2002, pp. 224 - 255, available at: http://www.freidok.uni-freiburg.de/volltexte/5029/pdf/Verfassungsrecht_der_Humangenetik_im_deutsch_japanischen_Vergleich.pdf

Nettesheim, Martin, "Grund und Grenzen der Wissenschaftsfreiheit", *Deutsches Verwaltungsblatt* 17 (2005), pp. 1072 - 1081

Nettesheim, Martin, "Kompetenzen", in: Armin von Bogdandy (ed.), *Europäisches Verfassungsrecht*, Springer, Berlin, 1st ed. 2003, pp. 415 - 477

Neumeyer, Karl, *Internationales Verwaltungsrecht, Vol. IV, Allgemeiner Teil,* Schweitzer, München 1936

Neuschwander, Thomas, *Mythos MITI*: Ind*ustriepolitik in Japan,* Lang, Frankfurt am Main/Berlin 1994

Newman, William, *Atoms and Alchemy. Chemistry and the Experimental Origins of Scientific Revolution,* University of Chicago Press, Chicago 2006

Nordquist, Myron H./Alexander Yankov/Neal R. Grandy/Shabtai Rosenne, *United Nations Convention on the Law of the Sea 1982 – A Commentary,* Vol. IV, Nijhoff, Dordrecht 2002

Nowak, Manfred, *U.N. Covenant on Civil and Political Rights. CCPR Commentary,* Engel, Kehl/Straßbourg/Arlington, 2nd ed. 2005

Nowotny, Helga/Peter Scott/Michael Gibbons, *Re-Thinking Science: Knowledge and the Public in an Age of Uncertainty,* Polity Press, Cambridge 2001

Nowotny, Helga, "The Changing Nature of Public Science", in: id./ Dominique Pestre/Eberhard Schmidt-Aßmann/Helmuth Schulze-Fielitz/Hans-Heinrich Trute, *The Public Nature of Science under Assault,* Springer, Berlin 2005, pp. 1 - 27

Ohler, Christoph, *Die Kollisionsordnung des Allgemeinen Verwaltungsrechts,* Mohr Siebeck, Tübingen 2005

Oppermann, Thomas, "Freiheit von Forschung und Lehre", in: Josef Isensee and Paul Kirchhof (eds.), *Handbuch des Staatsrechts, Band VI Freiheitsrechte*, Müller, Heidelberg 2001, pp. 809 - 845

Ornstein, Marta, *The Role of Scientific Societies in the Seventeenth Century,* University of Chicago Press, Chicago 1975

Pähler, Klaus, *Qualitätsmerkmale wissenschaftlicher Theorien,* Mohr Siebeck, Tübingen 1986

Park Si-soo, "Hwang Convicted of Embezzlement, Cleared of Fraud", published online 26 October 2006, *The Korea Times*, available at http://www.koreatimes.co.kr/www/news/nation/2009/10/117_54275.html

Pasteur, Louis, *Inauguration de l'Institut Pasteur, Annales de l'Institut Pasteur*, Masson, Paris 1888

Paulus, Andreas, *Die internationale Gemeinschaft im Völkerrecht,* Beck, München 2001

Pauss, Felicitas, "Das CERN: Beispiel einer erfolgreichen europäischen Forschungsinstitution", in: Gerhart von Graevenitz and Jürgen Mittelstraß (eds.), *Das Projekt Europa: Rechts-, forschungs- und kulturpolitische Aspekte*, UVK Universitätsverlag Konstanz 2009, pp. 89-98

Pellet, Alain, "Article 38", in: Andreas Zimmermann/Christian Tomuschat/Karin Oellers-Frahm (eds.), *The Statute of the International Court of Justice,* Oxford University Press, Oxford 2006, pp. 677 - 792

Pernice, Ingolf, "The Global Dimension of Multilevel Constitutionalism: A Legal Response to the Challenges of Globalisation", in: Pierre-Marie Dupuy/Bardo Fassbender/Malcolm N. Shaw/Karl-Peter Sommermann (eds.), *Völkerrecht als Wertordnung: Festschrift für Christian Tomuschat,* Engel, Kehl 2006, pp. 973 - 1005

Pestre, Dominique, "The Technosciences between Markets, Social Worries and the Political: How to Imagine a Better Future", in: Helga Nowotny/Dominique Pestre/Eberhard Schmidt-Aßmann/Helmuth Schulze-Fielitz/Hans-Heinrich Trute, *The Public Nature of Science under Assault,* Springer, Berlin 2005, pp. 29 - 52

Peters, Anne, "Global Constitutionalism in a Nutshell", in: Klaus Dicke/Stephan Hobe/Karl-Ulrich Meyn/id./Eibe Riedel/Hans-Joachim Schütz/Christian Tietje (eds.), *Weltinnenrecht,* Duncker & Humboldt, Berlin 2005, pp. 535 - 550

Petersmann, Ernst-Ulrich, *Constitutional functions and constitutional problems of international economic law,* University Press Fribourg Switzerland, Fribourg 1991

Pichot, André, *Die Geburt der Wissenschaft. Von den Babyloniern zu den frühen Griechen,* Campus-Verlag, Frankfurt 1995

Plesmann, Wolf and Volker Röben, "Marine Scientific Research: State Practice versus Law of the Sea?", in: Rüdiger Wolfrum (ed.) *Law of the sea at the crossroads: the continuing search for a universally accepted régime*, Duncker & Humboldt, Berlin 1991, pp. 373 - 392

Radermacher, Hans, *Cartesianische Wissenschaftstheorie,* Heiderhoff, Frankfurt am Main 1971

Rajagopal, Balakrishnan, "Academic Freedom as a Human Right. An Internationalist Perspective", in: *Academe* Vol. 89, issue 3 (May - June 2003), pp. 25 - 28

Raustiala, Kal, "The Architecture of International Cooperation: Transgovernmental Networks and the Future of International Law, *Virginia Journal of International Law* 43 (2001), pp. 1 - 92

Regent, Sabrina, "The Open Method of Coordination: A New Supranational Form of Governance?", *European Law Journal* 9 (2003), pp. 190 - 214

Reuter, Thomas, *Die ESA als Raumfahrtagentur der Europäischen Union*, Heymanns, Köln 2007

Riedel, Eibe, "Global Responsibilities and Bioethics: Reflections on the Council of Europe's Bioethics Convention", *Indiana Journal of Global Legal Studies* 5 (1997), pp. 179-190

Rittberger, Volker and Bernhard Zangl, *Internationale Organisationen*, Leske + Budrich, Opladen, 3rd ed. 2003

Röben, Volker, "The Sciences – A Contribution to Understanding the Law on an Activity of International Concern", *German Yearbook of International Law* 37 (1994), pp. 254 - 280

Röhrs, Hermann, *Der Einfluss der klassischen deutschen Universitätsidee auf die Higher Education in Amerika*, Deutscher Studienverlag, Weinheim 1995

Rosenau James N. and Ernst Otto Czempiel (eds.), *Governance without Government,* Cambridge University Press, Cambridge 1992

Rosenau, James, *Along the Domestic-Foreign Frontier – Exploring Governance in a Turbulent World,* Cambridge University Press, Cambridge 1997

Ruffert, Matthias, "Grund und Grenzen der Wissenschaftsfreiheit", *Veröffentlichungen der Vereinigung der Deutschen Staatsrechtslehrer* 65 (2006), pp. 146 - 210

Ruffert, Matthias, "Perspektiven des Internationalen Verwaltungsrechts", in: Christoph Möllers/Andreas Voßkuhle/Christian Walter (eds.), *Internationales Verwaltungsrecht,* Mohr Siebeck, Tübingen 2007, pp. 395 - 420

Ruffert, Matthias, *Die Globalisierung als Herausforderung an das Öffentliche Recht,* Boorberg, Stuttgart 2004

Ruffert, Matthias, in: Christian Calliess/id. (eds.), *EUV/AEUV*, Beck, München, 4th ed. 2011

Ruffert, Matthias, "The Transformation of Administrative Law as a Transnational Methodological Project", in: id. (ed.), *The Transformation of Administrative Law in Europe – La mutation du droit administrative en Europe*, Sellier, München 2007, pp. 3 - 52

Ruffert, Matthias, "Das Umweltvölkerrecht im Spiegel der Erklärung von Rio und der Agenda 21", *Zeitschrift für Umweltrecht* 1993, pp. 208 - 214

Ruffert, Matthias, "Von der Europäisierung des Verwaltungsrechts zum Europäischen Verwaltungsverbund, *Die Öffentliche Verwaltung* 2007, pp. 761 - 770

Ruffert, Matthias, "Verselbständigte Verwaltungseinheiten: Ein europäischer Megatrend im Vergleich", in: Hans-Heinrich Trute/Thomas Groß/Hans Christian Röhl/Christoph Möllers (eds.), *Allgemeines Verwaltungsrecht – zur Tragfähigkeit eines Konzepts*, Mohr Siebeck, Tübingen 2008, pp. 431 - 457

Ruffert, Matthias, "Industriepolitik: Staatsdirigismus in der Marktwirtschaft?", in: Jürgen F. Baur/Otto Sandrock/Boris Scholtka/Amos Shapira (eds.), *Festschrift für Gunther Kühne zum 70. Geburtstag*, Verlag Recht und Wirtschaft, Frankfurt am Main 2009, pp. 1021 - 1036

Ruffert, Matthias, "Rechtsquellen und Rechtsschichten des Verwaltungsrechts", in: Wolfgang Hoffmann-Riem/Eberhard Schmidt-Aßmann/Andreas Voßkuhle (eds.), *Grundlagen des Verwaltungsrechts, Band I*, Beck, München 2006, pp. 1085 - 1173

Russo, Lucio, *The Forgotten Revolution*, Springer, Berlin 2004

Saladin, Peter, *Wozu noch Staaten?: zu den Funktionen eines modernen demokratischen Rechtsstaats in einer zunehmend überstaatlichen Welt*, Stämpfli, Bern 1995

Sands, Philippe and Pierre Klein, *Bowett's Law of International Institutions*, Sweet & Maxwell, London, 6th ed. 2009

Schelsky, Helmut, *Einsamkeit und Freiheit. Idee und Gestalt der deutschen Universität und ihrer Reformen*, Rowohlt, Rheinbeck bei Hamburg 1963

Schiavone, Guiseppe, *International Organizations – A Dictionary and Directory*, Palgrave Macmillan, Houndmills/New York, 7th ed. 2008

Schiller, Friedrich (together with Johann Wolfgang von Goethe), "Xenien aus dem Musen-Almanach für das Jahr 1797", in: Friedrich Schiller, *Gedichte* (ed. by Georg Kurscheidt), Deutscher Klassiker Verlag, Frankfurt am Main 1992, pp. 577 - 629

Schlemmer-Schulte, Sabine, "Internationales Währungs- und Finanzrecht", in: Christian Tietje (ed.), *Internationales Wirtschaftsrecht,* Berlin/New York 2009, pp. 377 - 433

Schmidt, Harald, "Bioethics, Human Rights and Universalisation: a Troubled Relationship? – Observations on UNESCO's Universal Declaration on Bioethics and Human Rights", in: Silja Vöneky/Cornelia Hagedorn/Miriam Clados/Jelena von Achenbach (eds.), *Legitimation ethischer Entscheidungen im Recht,* Springer, Berlin 2009, pp. 275 - 295

Schmidt-Aßmann, Eberhard, "The Internationalization of Administrative Relations as a Challenge for Administrative Law Scholarship", in: Armin von Bogdandy/Rüdiger Wolfrum/Jochen von Bernstorff/Philipp Dann/Matthias Goldmann (eds.), *The Exercise of Public Authority by International Institutions*, Springer, Heidelberg 2010, pp. 943 - 964

Schmidt-Aßmann, Eberhard, in: id. and Bettina Schöndorf-Haubold (eds.), *Der Europäische Verwaltungsverbund*, Mohr Siebeck, Tübingen 2005, pp. 1 - 23

Schmidt-Aßmann, Eberhard, *Das allgemeine Verwaltungsrecht als Ordnungsidee*, Springer, Berlin, 2nd ed. 2004

Schneider, Hans Joachim, *Kriminologie für das 21. Jahrhundert – Schwerpunkt und Fortschritte der internationalen Kriminologie*, LIT-Verlag, Münster 2001

Schulze-Fielitz, Helmuth, "Die Verwaltung im europäischen Verfassungsgefüge", in: Wilfried Erbguth and Johannes Masing (eds.), *Verwaltung unter dem Einfluß des Europarechts*, Boorberg, Stuttgart 2006, pp. 91 - 136

Schulze-Fielitz, Helmuth, "Politische Voraussetzungen wissenschaftlicher Forschung", in: Horst Dreier and Dietmar Willoweit (eds.), *Wissenschaft und Politik,* Steiner, Stuttgart 2010, pp. 71 - 106

Schuppert, Gunnar Folke and Michael Zürn (eds.), "Governance in einer sich wandelnden Welt", *Politische Vierteljahresschrift-Sonderheft*, Springer, Heidelberg 41/2008

Schuppert, Gunnar Folke (ed.), *Governance als Prozess,* Nomos, Baden-Baden 2009

Schuppert, Gunnar Folke, "Was ist und wozu Governance", *Die Verwaltung* 40 (2007), pp. 461 - 511

Schwinges, Christoph, *Humboldt International. Der Export des deutschen Universitätsmodells im 19. und 20. Jahrhundert*, Schwabe, Basel 2001

Slaughter, Anne-Marie, *A New World Order,* Princeton University Press, Princeton 2004

Smend, Rudolf, "Das Recht der freien Meinungsäußerung", *Veröffentlichungen der Vereinigung der Deutschen Staatsrechtslehrer* 4 (1928), pp. 44 - 74

Smolentseva, Anna, "Challenges to the Russian academic profession", *Higher Education* 45 (2003), pp. 391 - 424

Standler, Ronald, *Academic Freedom in the USA*, 1999, available at www.rbs2.com/afree.htm

Starck, Christian, "Article 5", in: Hermann von Mangoldt/Friedrich Klein/id. (eds.), *Kommentar zum Grundgesetz*, Vol. 1, Vahlen, München, 6th ed. 2010

Stehr, Nico, "Modern societies as knowledge societies", in: George Ritzer and Barry Smart (eds.), *Handbook of social theory,* Sage, London 2001, pp. 494 - 508

Stehr, Nico, *Wissen und Wirtschaften. Die gesellschaftlichen Grundlagen der modernen Ökonomie*, Suhrkamp, Frankfurt am Main 2001

Steinbrook, Robert, "Egg Donation and Human Embryonic Stem-Cell Research", *The New England Journal of Medicine* 354 (4), 26 January 2006, pp. 324 - 326

Steinecke, Sebastian, *Zur internationalen Governance der Wissenschaft,* 2010, manuscript, (forthcoming Herbert Utz, München, March 2011)

Stichweh, Rudolf, "The Multiple Publics of Science: Inclusion and Popularization", *Soziale Systeme* 9 (2003), pp. 210 - 220

Stoll, Peter-Tobias, "Freihandel und Verfassung", *Zeitschrift für ausländisches öffentliches Recht und Völkerrecht* 57 (1997), pp. 83 - 146

Straßburger, Jana, "Die Inkorporation der Deklaration von Helsinki in das ärztliche Berufs- und Standesrecht – Verfassungsrechtliche Aspekte", *Medizinrecht* 2006, pp. 462 - 471

Sydow, Gernot, "Die Vereinheitlichung des mitgliedstaatlichen Vollzugs des Europarechts in mehrstufigen Verwaltungsverfahren", *Die Verwaltung* 34 (2001), pp. 517 - 542

Szaif, Jan, "Die Geschichte des Wahrheitsbegriffs in der klassischen Antike", in: Markus Enders and id. (eds.), *Die Geschichte des philosophischen Begriffs der Wahrheit*, de Gruyter, Berlin 2006, pp. 1 - 32

Talmon, Stefan, "The Security Council as World Legislature", *The American Journal of International Law* 99 (2005), pp. 175 - 193

Thagard, Paul, "Why astrology is Pseudoscience", *Proceedings of the Biennial Meeting of the Philosophy of Science Association* 1 (1978), p. 223 - 234

The Senate. Standing Committee on Education, Employment and Workplace Relations (ed.), *Report Allegations of Academic Bias in Universities and Schools,* 2008, available at http://www.aph.gov.au/senate/committee/eet_ctte/academic_freedom/report/report.pdf

Theobald, Christian, "Die Weltbank: Good Governance und die Neue Institutionenökonomik", *Verwaltungsarchiv* 89 (1999), pp. 467 - 487

Thomson, James A. et al., "Embryonic Stem Cell Lines Derived from Human Blastocysts", *Science* 282 (1998), pp. 1145 - 1147

Tietje, Christian, *Internationalisiertes Verwaltungshandeln,* Duncker & Humboldt, Berlin 2001

Tietje, Christian/Alan Brouder/Karsten Nowrot, *Philip C. Jessup's Transnational Law Revisited – On the Occasion of the 50th Anniversary of its Publication,* Universität Halle-Wittenberg, Halle 2006

Tietje, Christian, "The Changing Legal Structure of International Treaties as an Aspect of an Emerging Global Governance Architecture", *German Yearbook of International Law* 42 (2000), pp. 26 - 55

Tobey, Ronald, *The American Ideology of National Science 1919–1930,* University of Pittsburgh Press, Pittsburgh 1971

Tomuschat, Christian, "Die internationale Gemeinschaft", *Archiv des Völkerrechts* 33 (1995), pp. 1 - 20

Trute, Hans-Heinrich/Doris Kühlers/Arne Pilniok, "Der Governance-Ansatz als verwaltungsrechtswissenschaftliches Analysekonzept", in: Gunnar Folke Schuppert and Michael Zürn (eds.), Governance in einer sich wandelnden Welt, *Politische Vierteljahresschrift-Sonderheft 41/2008*, Springer, Heidelberg, pp. 173 - 189

Trute, Hans-Heinrich, *Die Forschung zwischen grundrechtlicher Freiheit und staatlicher Institutionalisierung. Das Wissenschaftsrecht als Recht kooperativer Verwaltungsvorgänge*, Mohr Siebeck, Tübingen 1994

Turner, Roy, "Humboldt in North America? Reflections on the Research University and its Historians", in: Christoph Schwinges, *Humboldt International. Der Export des deutschen Universitätsmodells im 19. und 20. Jahrhundert*, Schwabe, Basel 2001, pp. 289 - 312

Uerpmann, Robert, "Mittelbare Gemeinschaftsverwaltung durch gemeinschaftsgeschaffene juristische Personen des öffentlichen Rechts", *Archiv des öffentlichen Rechts* 125 (2000), pp. 551 - 586

Verdross, Alfred, *Die Verfassung der Völkerrechtsgemeinschaft,* Springer, Wien 1926

Verdross, Alfred and Bruno Simma, *Universelles Völkerrecht,* Duncker & Humboldt, Berlin, 3rd ed. 1984

Vitanyi, Béla, "Les positions doctrinales concernant le sens de la notion de 'principes généraux de droit reconnus par les nations civilisées'", *Revue Générale de Droit International Public* 86 (1), 1982, pp. 48-116

Vöneky, Silja, "Ethische Experten und moralischer Autoritarismus", in: id./Cornelia Hagedorn/Miriam Clados/Jelena von Achenbach (eds.), *Legitimation ethischer Entscheidungen im Recht,* Springer, Berlin 2009, pp. 85 - 97

Vöneky, Silja and Rüdiger Wolfrum (eds.), *Human Dignity and Human Cloning,* Nijhoff, Leiden 2004

Vogel, Klaus, *Der räumliche Anwendungsbereich der Verwaltungsrechtsnorm,* Metzner, Frankfurt am Main/Berlin 1965

Vos, Ellen, "Reforming the European Commission: What Role to Play for European Agencies", *Common Market Law Review* 37 (2000), pp. 1113 - 1134

Wächter, Bernd, *Handbook of European Associations in Higher Education: A Practical Guide to Academic Networks in Europe and Beyond*, Lemmens, Bonn 2000

Wagner, Helmut, "Gibt es ein Grundrecht der Wissenschaftsfreiheit im Europäischen Gemeinschaftsrecht", *Die öffentliche Verwaltung* 1999, pp. 129 - 137

Wahl, Rainer, "Konstitutionalisierung – Leitbegriff oder Allerweltsbegriff?", in: Carl-Eugen Eberle (ed.), *Festschrift Winfried Brohm,* Beck, München 2002, pp. 191 - 207

Wallis, Roy (ed.), *On the Margins of Science: The Social Construction of Received Knowledge,* University of Keele Press, Edinburgh 1979

Walter, Christian and Matthias Ruffert, *Institutionalisiertes Völkerrecht,* Beck, München 2009

Walterscheid, Edward C., "To Promote the Progress of Science and Useful Arts: The anatomy of a Congressional Power", *The Journal of Law and Technology* 43 (2002), pp. 1 - 81

Walterscheid, Edward C., "The Preambular Argument: the Dubious Premise of Eldred v. Ashcroft", *The Journal of Law and Technology* 44 (2004), pp. 331 - 379

Weil, Prosper, "Towards Relative Normativity in International Law?", *American Journal of International Law* 1983, pp. 413 - 442

Weiler, Joseph H. H., "The Geology of International Law – Governance, Democracy and Legitimacy", *Zeitschrift für ausländisches öffentliches Recht und Völkerrecht* 64 (2004), pp. 547 - 562

Weingart, Peter, *Wissenschaftssoziologie,* Transcript-Verlag, Bielefeld 2003

Weiß, Wolfgang, *Der Europäische Verwaltungsverbund*, Duncker & Humboldt, Berlin 2010

Wihl, Tim, "Freiheit als Unwert? Verwandlungen des Völkerrechts aus liberaler Perspektive", in: Christian Tomuschat (ed.), *Weltordnungsmodelle für das 21. Jahrhundert,* Nomos, Baden-Baden 2009, pp. 65 - 98

Willke, Helmut, *Dystopia*, Suhrkamp, Frankfurt am Main 2002

Winter, Alison, *Mesmerized. Powers of Mind in Victorian Britain,* University of Chicago Press, Chicago 1998

Wyatt, Derrick/Alan Dashwood/Michael Dougan/Anthony Arnull/Malcolm Ross/Elanor Spaventa, *Wyatt & Dashwood's European Union Law,* Sweet & Maxwell, London, 5th ed. 2006

Zacharias, Diana, "The UNESCO Regime for the Protection of World Heritage as Prototype of an Autonomy-Gaining International Institution", in: Armin von Bogdandy/Rüdiger Wolfrum/Jochen von Bernstorff/Philipp Dann/Matthias Goldmann (eds.), *The Exercise of Public Authority by International Institutions*, Springer, Heidelberg 2010, pp. 301 - 336

Zuckerman, Harriet, "Sociology of Science", in: Neil Joseph Smelser (ed.), *Handbook of Sociology*, Sage, Newbury Park 1988, pp. 511 - 574

Max-Planck-Institut für ausländisches öffentliches Recht und Völkerrecht

Beiträge zum ausländischen öffentlichen Recht und Völkerrecht

Hrsg.: A. von Bogdandy, R. Wolfrum

Bde. 27–59 erschienen im Carl Heymanns Verlag KG Köln, Berlin (Bestellung an: Max-Planck-Institut für Völkerrecht, Im Neuenheimer Feld 535, 69120 Heidelberg); ab Band 60 im Springer-Verlag Berlin, Heidelberg, New York, London, Paris, Tokyo, Hong Kong, Barcelona

---

228 Matthias *Ruffert*, Sebastian *Steinecke*: **The Global Administrative Law of Science.** 2011. IX, 140 Seiten. Geb. € 59,95

227 Sebastian *Pritzkow*: **Das völkerrechtliche Verhältnis zwischen der EU und Russland im Energiesektor.** 2011. XXIV, 304 Seiten. Geb. € 79,95

225 Clemens *Feinäugle*: **Hoheitsgewalt im Völkerrecht.** 2011. XXVI, 418 Seiten. Geb. € 89,95

224 David *Barthel*: **Die neue Sicherheits- und Verteidigungsarchitektur der Afrikanischen Union.** 2011. XXV, 443 Seiten. Geb. € 94,95

223 Tilmann *Altwicker*: **Menschenrechtlicher Gleichheitsschutz.** 2011. XXX, 549 Seiten. Geb. € 99,95

222 Stephan *Bitter*: **Die Sanktion im Recht der Europäischen Union.** 2011. XV, 351 Seiten. Geb. € 84,95

221 Holger *Hestermeyer*, Nele *Matz-Lück*, Anja *Seibert-Fohr*, Silja *Vöneky* (eds.): **Law of the Sea in Dialogue.** 2011. XII, 189 Seiten. Geb. € 69,95 zzgl. landesüblicher MwSt.

220 Jan *Scheffler*: **Die Europäische Union als rechtlich-institutioneller Akteur im System der Vereinten Nationen.** 2011. XXXV, 918 Seiten. Geb. € 149,95

219 Mehrdad *Payandeh*: **Internationales Gemeinschaftsrecht.** 2010. XXXV, 629 Seiten. Geb. € 99,95

218 Jakob *Pichon*: **Internationaler Strafgerichtshof und Sicherheitsrat der Vereinten Nationen.** 2011. XXVI, 399 Seiten. Geb. € 89,95

217 Michael *Duchstein*: **Das internationale Benchmarkingverfahren und seine Bedeutung für den gewerblichen Rechtsschutz.** 2010. XXVI, 528 Seiten. Geb. € 99,95

216 Tobias *Darge*: **Kriegsverbrechen im nationalen und internationalen Recht.** 2010. XXXV, 499 Seiten. Geb. € 94,95

215 Markus *Benzing*: **Das Beweisrecht vor internationalen Gerichten und Schiedsgerichten in zwischenstaatlichen Streitigkeiten.** 2010. L, 846 Seiten. Geb. € 139,95

214 Urs *Saxer*: **Die internationale Steuerung der Selbstbestimmung und der Staatsentstehung.** 2010. XLII, 1140 Seiten. Geb. € 169,95

213 Rüdiger *Wolfrum*, Chie *Kojima* (eds.): **Solidarity: A Structural Principle of International Law.** 2010. XIII, 238 Seiten. Geb. € 69,95

212 Ramin S. *Moschtaghi*: **Die menschenrechtliche Situation sunnitischer Kurden in der Islamischen Republik Iran.** 2010. XXIII, 451 Seiten. Geb. € 94,95

211 Georg *Nolte* (ed.): **Peace through International Law. The Role of the International Law Commission.** 2009. IX, 195 Seiten. Geb. € 64,95 zzgl. landesüblicher MwSt.

210 Armin *von Bogdandy*, Rüdiger *Wolfrum*, Jochen *von Bernstorff*, Philipp *Dann*, Matthias *Goldmann* (eds.): **The Exercise of Public Authority by International Institutions.** 2010. XIII, 1005 Seiten. Geb. € 149,95 zzgl. landesüblicher MwSt.

209 Norman *Weiß*: **Kompetenzlehre internationaler Organisationen.** 2009. XVIII, 540 Seiten. Geb. € 99,95

208 Michael *Rötting*: **Das verfassungsrechtliche Beitrittsverfahren zur Europäischen Union.** 2009. XIV, 317 Seiten. Geb. € 79,95

207 Björn *Ahl*: **Die Anwendung völkerrechtlicher Verträge in China.** 2009. XIX, 419 Seiten. Geb. € 89,95

206 Mahulena *Hofmann*: **Von der Transformation zur Kooperationsoffenheit?** 2009. XIX, 585 Seiten. Geb. € 99,95

205 Rüdiger *Wolfrum*, Ulrike *Deutsch* (eds.): **The European Court of Human Rights Overwhelmed by Applications: Problems and Possible Solutions.** 2009. VIII, 128 Seiten. Geb. € 59,95 zzgl. landesüblicher MwSt.

204 Niels *Petersen*: **Demokratie als teleologisches Prinzip.** 2009. XXVII, 280 Seiten. Geb. € 79,95

203 Christiane *Kamardi*: **Die Ausformung einer Prozessordnung sui generis durch das ICTY unter Berücksichtigung des Fair-Trial-Prinzips.** 2009. XVI, 424 Seiten. Geb. € 89,95

202 Leonie F. *Guder*: **The Administration of Debt Relief by the International Financial Institutions.** 2009. XVIII, 355 Seiten. Geb. € 84,95 zzgl. landesüblicher MwSt.

201 Silja *Vöneky*, Cornelia *Hagedorn*, Miriam *Clados*, Jelena *von Achenbach*: **Legitimation ethischer Entscheidungen im Recht.** 2009. VIII, 351 Seiten. Geb. € 84,95

200 Anja Katarina *Weilert*: **Grundlagen und Grenzen des Folterverbotes in verschiedenen Rechtskreisen.** 2009. XXX, 474 Seiten. Geb. € 94,95

199 Suzette V. *Suarez*: **The Outer Limits of the Continental Shelf.** 2008. XVIII, 276 Seiten. Geb. € 79,95 zzgl. landesüblicher MwSt.

198 Felix *Hanschmann*: **Der Begriff der Homogenität in der Verfassungslehre und Europarechtswissenschaft.** 2008. XIII, 370 Seiten. Geb. € 84,95

197 Angela *Paul*: **Kritische Analyse und Reformvorschlag zu Art. II Genozidkonvention.** 2008. XVI, 379 Seiten. Geb. € 84,95

196 Hans Fabian *Kiderlen*: **Von Triest nach Osttimor.** 2008. XXVI, 526 Seiten. Geb. € 94,95

195 Heiko *Sauer*: **Jurisdiktionskonflikte in Mehrebenensystemen.** 2008. XXXVIII, 605 Seiten. Geb. € 99,95

194 Rüdiger *Wolfrum*, Volker *Röben* (eds.): **Legitimacy in International Law.** 2008. VI, 420 Seiten. Geb. € 84,95 zzgl. landesüblicher MwSt.

193 Doris *König*, Peter-Tobias *Stoll*, Volker *Röben*, Nele *Matz-Lück* (eds.): **International Law Today: New Challenges and the Need for Reform?** 2008. VIII, 260 Seiten. Geb. € 69,95 zzgl. landesüblicher MwSt.

192 Ingo *Niemann*: **Geistiges Eigentum in konkurrierenden völkerrechtlichen Vertragsordnungen.** 2008. XXV, 463 Seiten. Geb. € 94,95

191 Nicola *Wenzel*: **Das Spannungsverhältnis zwischen Gruppenschutz und Individualschutz im Völkerrecht.** 2008. XXXI, 646 Seiten. Geb. € 99,95

190 Winfried *Brugger*, Michael *Karayanni* (eds.): **Religion in the Public Sphere: A Comparative Analysis of German, Israeli, American and International Law.** 2007. XVI, 467 Seiten. Geb. € 89,95 zzgl. landesüblicher MwSt.

189 Eyal *Benvenisti*, Chaim *Gans*, Sari *Hanafi* (eds.): **Israel and the Palestinian Refugees.** 2007. VIII, 502 Seiten. Geb. € 94,95 zzgl. landesüblicher MwSt.

188 Eibe *Riedel*, Rüdiger *Wolfrum* (eds.): **Recent Trends in German and European Constitutional Law.** 2006. VII, 289 Seiten. Geb. € 74,95 zzgl. landesüblicher MwSt.

187 Marcel *Kau*: **United States Supreme Court und Bundesverfassungsgericht.** 2007. XXV, 538 Seiten. Geb. € 99,95 zzgl. landesüblicher MwSt.

186 Philipp *Dann*, Michał *Rynkowski* (eds.): **The Unity of the European Constitution.** 2006. IX, 394 Seiten. Geb. € 79,95 zzgl. landesüblicher MwSt.

185 Pál *Sonnevend*: **Eigentumsschutz und Sozialversicherung.** 2008. XVIII, 278 Seiten. Geb. € 74,95

184 Jürgen *Bast*: **Grundbegriffe der Handlungsformen der EU.** 2006. XXI, 485 Seiten. Geb. € 94,95

183 Uwe *Säuberlich*: **Die außervertragliche Haftung im Gemeinschaftsrecht.** 2005. XV, 314 Seiten. Geb. € 74,95

182 Florian *von Alemann*: **Die Handlungsform der interinstitutionellen Vereinbarung.** 2006. XVI, 518 Seiten. Geb. € 94,95

181 Susanne *Förster*: **Internationale Haftungsregeln für schädliche Folgewirkungen gentechnisch veränderter Organismen.** 2007. XXXVI, 421 Seiten. Geb. € 84,95

180 Jeanine *Bucherer*: **Die Vereinbarkeit von Militärgerichten mit dem Recht auf ein faires Verfahren gemäß Art. 6 Abs. 1 EMRK, Art. 8 Abs. 1 AMRK und Art. 14 Abs. 1 des UN-Paktes über bürgerliche und politische Rechte.** 2005. XVIII, 307 Seiten. Geb. € 74,95

179 Annette *Simon*: **UN-Schutzzonen – Ein Schutzinstrument für verfolgte Personen?** 2005. XXI, 322 Seiten. Geb. € 74,95

178 Petra *Minnerop*: **Paria-Staaten im Völkerrecht?** 2004. XXIII, 579 Seiten. Geb. € 99,95

177 Rüdiger *Wolfrum*, Volker *Röben* (eds.): **Developments of International Law in Treaty Making.** 2005. VIII, 632 Seiten. Geb. € 99,95 zzgl. landesüblicher MwSt.

176 Christiane *Höhn*: **Zwischen Menschenrechten und Konfliktprävention. Der Minderheitenschutz im Rahmen der Organisation für Sicherheit und Zusammenarbeit in Europa (OSZE).** 2005. XX, 418 Seiten. Geb. € 84,95

MIX
Papier aus verantwortungsvollen Quellen
Paper from responsible sources
FSC® C105338

If you have any concerns about our products, you can contact us on
ProductSafety@springernature.com

In case Publisher is established outside the EU, the EU authorized representative is:
Springer Nature Customer Service Center GmbH
Europaplatz 3, 69115 Heidelberg, Germany

Printed by Libri Plureos GmbH
in Hamburg, Germany